Space/Terrestrial Mobile Networks

Space/Terrestrial Mobile Networks
Internet Access and QoS Support

Edited by

R.E. Sheriff
University of Bradford, UK

Y.F. Hu
University of Bradford, UK

G. Losquadro
Alenia Spazio, Italy

P. Conforto
Alenia Spazio, Italy

C. Tocci
Alenia Spazio, Italy

John Wiley & Sons, Ltd

Other Wiley Editorial Offices

John Wiley & Sons Inc., 111 River Street, Hoboken, NJ 07030, USA

Jossey-Bass, 989 Market Street, San Francisco, CA 94103-1741, USA

Wiley-VCH Verlag GmbH, Boschstr. 12, D-69469 Weinheim, Germany

John Wiley & Sons Australia Ltd, 33 Park Road, Milton, Queensland 4064, Australia

John Wiley & Sons (Asia) Pte Ltd, 2 Clementi Loop #02-01, Jin Xing Distripark, Singapore 129809

John Wiley & Sons Canada Ltd, 22 Worcester Road, Etobicoke, Ontario, Canada M9W 1L1

Wiley also publishes its books in a variety of electronic formats. Some of the content that appears in
print may not be available in electronic books.

British Library Cataloguing in Publication Data

A catalogue record for this book is available from the British Library

0-470-85031-0

Typeset in 10/12pt Times by Thomson Press (India) Limited, New Delhi
Printed and bound in Great Britain by Antony Rowe Ltd, Chippenham, Wiltshire
This book is printed on acid-free paper responsibly manufactured from sustainable forestry
in which at least two trees are planted for each one used for paper production.

Contents

Preface

This book is based on the work that was performed as part of the European Union's Fifth Framework Information Society Technologies (IST) Programme, which spanned the period 1998–2002. Specifically, research and development was performed under the project SUITED: Multi-Segment System for Ubiquitous Access to Internet Services and Demonstrator, which was led by Alenia Spazio of Italy. This 30-month, 4 million Euro project brought together thirteen organisations from across Europe, each partner bringing to the consortium expertise in a particular area. In total, the SUITED project comprised of over 900 person-months of effort, generated 16 project deliverables, as well as numerous publications in academic Journals and Conferences, developed various hardware and software components, and culminated with a series of trials involving satellite and cellular technologies. This book aims to provide a unique insight into the work and achievements that were associated with the SUITED project. The book illustrates the design process from study phase through to demonstration, while at the same time introducing the various technologies and concepts that were associated with the project.

The basic premise of the SUITED project is that satellite, cellular and wireless local area network (W-LAN) access networks inter-work with each other and with the Internet core network to provide mobile Internet type services with a guaranteed end-to-end (E2E) quality of service (QoS). This multi-segment access network in combination with a federation of Internet Service Providers (F-ISP) comprises the Global Mobile Broadband System (GMBS).

In today's telecommunications environment, there are numerous ways of establishing a communications link while on the move, anywhere and at anytime. In Europe, third-generation Universal Mobile Telecommunications System (UMTS) services have recently been launched, which now operate alongside second-generation GSM (Global System for Mobile communications) and General Packet Radio Service (GPRS). The influence of W-LAN technologies, chiefly driven by the IEEE 802.11 family of standards, is rapidly becoming a preferred means of establishing broadband communications in 'hot spot' areas. In a similar way, the provision of broadband services via satellite, necessitating the need to move up in frequency from L-band (1.5/1.6 GHz) to Ka-band (20/30 GHz), promises to open up new market opportunities for next-generation satellite-mobile services. One such planned satellite system being EuroSkyWay (ESW), proposed by Alenia Spazio. In future-generation mobile networks, the ability to provide services of guaranteed quality is seen as a key requirement. To achieve E2E QoS requires that the Internet must be able to guarantee service provision at an acceptable level rather than the best effort approach currently implemented.

Chapter 1 introduces the communication infrastructure of the GMBS. After outlining the general objectives of the GMBS, the underlying technologies and ways of operation behind the F-ISP are presented as a basis for the move towards a QoS aware mobile Internet. The constituents of the F-ISP and their respective roles in providing a guaranteed QoS are discussed.

The GMBS comprises of the access segments Mobile-ESW (M-ESW), GPRS, UMTS and W-LAN. Each of these access segments are described in Chapter 2. Each access segment is presented in terms of the system architecture, the network nodes, the access, modulation and coding, QoS support where applicable and user and control planes.

The GMBS Multi-Mode Terminal (GMMT) provides users with connectivity to the segments that provide access to the GMBS. The design of the GMMT is presented in Chapter 3. Key to the operation of the GMMT is the Terminal Inter-Working Unit (T-IWU), which participates in the inter-segment mobility and QoS procedures. The various modules associated with the T-IWU are defined. The constituent Mobile Terminals, which provide connectivity to the various access segments are then described.

The service requirements of the GMBS are presented in Chapter 4. Initially, the service scenarios are discussed, illustrating the coverage afforded by M-ESW, and the services supported by the Internet. Issues surrounding the provision of QoS are then discussed for typical GMBS type services. Security requirements at the access and Internet networks then conclude the service requirements.

The means of providing E2E QoS is described in Chapter 5. Two novel approaches to this problem are presented. A Quality of Service Support Module (QASM), which resides in the Terminal Inter-Working Unit of the GMMT and the Gateway GPRS Support Nodes (GGSNs) of UMTS/GPRS and the FES of the M-ESW access networks, is proposed for ensuring, by co-operating with the access segments' specific mechanisms, QoS requirements and efficient bandwidth utilisation from the user up to the edge router of the F-ISP. The Gauge&Gate Reservation with Independent Probing (GRIP) is proposed for establishing guaranteed QoS over the core portion of the F-ISP. Both methods inter-work with the Resource Reservation Protocol (RSVP) which is adopted in the edge portion of the GMBS network.

Issues surrounding mobility support are presented in Chapter 6. The Chapter considers location management, address management, handover management and IP mobility support issues. A novel approach to performing inter-segment handover, also known as vertical handover, using a fuzzy logic approach concludes the Chapter.

Chapter 7 presents a methodology used to design the network protocols associated with the GMBS. The approach involves of identification of functional entities, information flows, functional model, functional architecture and network architecture. The Chapter concludes with an example of how protocols are specified using SDL. The design process is illustrated by using the GMBS Registration procedure as a case study.

Chapter 8 concludes the book with the results of the demonstrations, which are divided into two phases, reflecting the laboratory and field trials environments. Examples of the equipment developed for the demonstrator are also presented.

Lastly, the publications resulting from the SUITED project are listed in Appendix A.

R.E. Sheriff
Y.F. Hu
G. Losquadro
P. Conforto
C. Tocci
17 December 2003

ACKNOWLEDGEMENTS

The SUITED project was performed by the following organisations: CoRiTeL—Consorzio di Ricerca Sulle Telecomunicazioni (Italy), Deutsches Zentrum für Luft- und Raumfahrt e.V. (Germany), Etnoteam S.p.a. (Italy), Siemens Aktiengesellschaft Oesterreich (Austria), Space Software Italia S.p.a. (Italy), Telit Mobile Terminals S.p.a. (Italy), TTI Norte, S.L. (Spain), Smartmove (Belgium), University of Bradford (UK), Universita degli Studi di Roma "La Sapienza" (Italy), Universita di Roma "Tor Vergata" (Italy). The Editors gratefully acknowledge the contributions made by all partners to the success of the SUITED project. The Editors also wish to acknowledge the contribution of the European Commission's Project Officer for SUITED, Mr. Bernard Barani.

The texts extracted from the ITU material have been reproduced with the prior authorisation of the Union as copyright holder.

The sole responsibility for selecting extracts for reproduction lies with the beneficiaries of this authorisation alone and can in no way be attributed to the ITU.

The complete volumes of the ITU material, from which the texts reproduced are extracted, can be obtained from:

International Telecommunication Union
Sales and Marketing Service
Place des Nations – CH-1211 GENEVA 20 (Switzerland)
Telephone: +41 22 730 61 41 (English)
+41 22 730 61 42 (French)
+41 22 730 61 43 (Spanish)
Telex: 421 000 uit ch / Fax: +41 22 730 51 94
X.400: S=sales; P=itu; A=400net; C=ch
E-mail: sales@itu.int / http://www.itu.int/publications

The editors would also like to acknowledge the use of the following source material within the publication.

Figures 1.1 and 3.1. P. Conforto, C. Tocci, V. Schena, L. Secondiani, N. Blefari-Melazzi, P.M.L. Chan, F. Delli Priscoli: "End-to-End QoS and Global Mobility Management in an Integrated Satellite/Terrestrial Network", INt. J. o Satellite Communications and Networking, 22(1), January-February 2004, Wiley; 19–54.

Figures 8.25 and 8.31. G. Bianchi, N. Blefari-Melazzi, P.M.L. Chan, M. Holzbock, Y.F. Hu, A. Jahn, R.E. Sheriff: "Design and Validation of QoS Aware Mobile Internet Access Procedures for Heterogeneous Networks", Mobile Networks and Applications – Special Issue Personal Environments Mobility in Multi-Provider and Multi-Segment Networks, Kluwer Academic Press, 8, January 2003; 11–25.

Figures 6.7 and 6.9. P.M.L. Chan, R.E. Sheriff, Y.F. Hu, P. Conforto, C. Tocci: "Mobility Management Incorporating Fuzzy Logic for a Heterogeneous IP Environment", IEEE Communications – Focus Issue on Evolving to Seamless All IP Wireless/Mobile Networks, 39(12), December 2001; 42–51.

Figures 3.2 and 5.3. P. Conforto, C. Tocci, G. Losquadro, R.E. Sheriff, P.M.L. Chan, Y.F. Hu: "Ubiquitous Internet in an Satellite-Terrestrial Environment", IEEE Communications – Focus Issue on Service Portability and Virtual Home Environment, 40(1), January 2002; 98–107.

1

Introduction

PAOLO CONFORTO, CLEMENTINA TOCCI
Alenia Spazio, Italy

1.1 GLOBAL MOBILE BROADBAND SYSTEM

The communication infrastructure of the *Global Mobile Broadband System* (GMBS) is described in this Chapter.

The network nodes forming the overall architecture are identified along with their mutual interconnections, the focus being on topological issues. A description of the functionality implemented in each node to support end-to-end (E2E) Quality of Service (QoS) and mobility will be given in Chapters 5 and 6, respectively.

The GMBS solution foresees a *multi-segment access network* consisting of broadband satellite [CAR-99] and wireless terrestrial components presenting mutually complementary features, which inter-work with the *Internet network*.

The final objective is represented by the creation of a GMBS based on an integrated satellite/terrestrial infrastructure.

From a user's perspective, the GMBS is perceived as a single network, capable of providing *"anywhere"*, *"anytime"* and *"anyhow"* QoS-guaranteed Internet services.

The task of GMBS is even more challenging when considering that it aims to support, not only the *best effort* (BE) services currently available over the Internet, but also *QoS-guaranteed* services.

The requirement of the GMBS to appear as a sole global network can only be achieved by fully merging each of the different components. In practice such a solution would be unrealisable, since any modifications to existing standardised systems (General Packet Radio Service (GPRS); Universal Mobile Telecommunications System (UMTS)) would be highly impractical and unlikely to find favour with operators or standards bodies alike. The followed approach foresees that the modifications to the operations of both the access and Internet networks composing the system should be kept to a minimum.

Space/Terrestrial Mobile Networks. Edited by R.E. Sheriff, Y.F. Hu, G. Losquadro, P. Conforto, C. Tocci
© 2004 John Wiley & Sons, Ltd ISBN: 0-470-85031-0

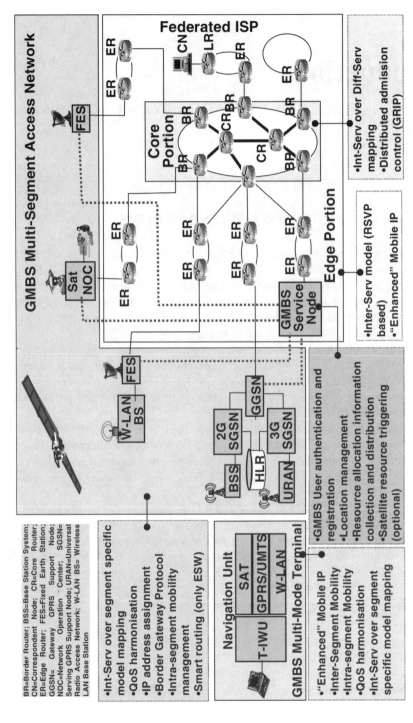

Figure 1.1 GMBS architecture. Reproduced by permission from John Wiley and Sons Inc. © 2004.

In order to meet the above requirements, the selected methodology foresees a step-by-step activity consisting of the following phases:

1. Identification of commonalities in the way of working of the access networks in order to create a possible parallel between network entities, terminal architecture, user/terminal identities, mobility management procedures and so on.

2. Design of *GMBS specific devices* and *procedures* to harmonise the co-operation and interaction among the segment specific procedures.

3. Implementation of minor modifications, only if strictly necessary, to the segment specific devices and procedures to complete the harmonisation between the respective networks.

The service area of the GMBS is considered to be world-wide and available in highly diversified environments, including:

- open/rural regions;
- suburban/urban areas;
- indoor and low-range outdoor localities.

Neither wireless terrestrial networks nor satellite systems operating independently are able to serve such a wide range of operating environments. In order to overcome this issue, the GMBS solution foresees a *multi-segment access network*, whose components offer complementary features. This heterogeneous network inter-works with the *Internet network*. From a user perspective, the GMBS is perceived as a single network that is capable of supporting QoS-guaranteed, Internet services when operating mobile or portable terminals.

An evolutionary scenario for GMBS deployment is envisaged, comprising of two major phases of development. In GMBS Phase I, the satellite segment is complemented by the GPRS system. While for GMBS Phase II, the UMTS segment will complement or replace the GPRS. In both phases also a wireless-LAN (W-LAN) segment is envisaged in order to bridge the satellite connectivity in some particular environments where the satellite coverage is not available. The segment availability per service rate in the different service environments is discussed in Chapter 3.

An evolutionary scenario is also envisaged for the Internet network within the GMBS. In the initial phase, Mobile IPv6 and QoS support functionality, which are required for the appropriate integration with GMBS, are assumed to be present only in a bounded portion of the overall Internet, referred to as the *federated Internet Service Provider* (F-ISP) *network*. Subsequent phases of implementation would result in the extension of the F-ISP network functionality to the remainder of the Internet. The overall GMBS architecture is depicted in Figure 1.1.

1.2 FEDERATED INTERNET SERVICE PROVIDER NETWORK

1.2.1 Rationale

Today's Internet consists of numerous Internet Service Providers (ISPs), each serving many constituent networks and end-users. The interaction of different ISPs takes place at both a technological level and a business model level. The former requires interconnection mechanisms (to handle routing signalling, traffic flows, etc.) usually named *"peering"*, while the latter refers to business arrangements also called *"settlements"* [HUS-99].

Currently, when networks adopt a BE service model, interactions among ISPs are defined by means of a *Service Level Agreement* (SLA). An SLA is a service contract that specifies how a given sub-network must handle the traffic that it receives from another sub-network upstream. The typical SLA that is currently stipulated specifies only two QoS parameters, i.e. bandwidth and availability. The contract is based on:

- *Bandwidth Guarantee:* the ISP provides the client with a guarantee that the network will not be a bottleneck for their communications. In particular, in a typical contract the ISP guarantees a minimum bandwidth of 95% of the nominal speed of the access port.

- *Availability Guarantee:* the ISP guarantees to the client service availability for a minimum of 99.9% of the contractual period.

Even though most commercial routers support both Differentiated-Services (DiffServ) [IET-98], for example priority queuing, weighted fair queuing, custom queuing and Integrated-Services (IntServ) [IET-94], for example Resource Reservation Protocol (RSVP) mechanisms, only a few ISPs exploit these functionalities in a commercial environment. An SLA defined in terms of bandwidth and availability guarantees is only possible in BE networks due to the fact that the BE model, neglecting other important elements (e.g. packet delay, jitter, throughput and loss), does not provide any mechanism to guarantee QoS. The only way to respect an SLA is to over-dimension the network links.

As far as mobility is concerned, most ISPs do not implement any specific mechanism to facilitate mobile usage. Moreover, most current mobile users connect to a single access point (typically a company access point or an ISP), thus they do not need any specific mobile IP (Internet Protocol) features. This restriction becomes unacceptable in a fully mobile environment, in which users are free to move over national boundaries.

Today's Internet is unable to meet the emerging needs of new categories of users, in terms of mobility and QoS guarantees. Since it is not possible to suddenly implement the functionality required to manage QoS and mobility over the whole Internet, an evolutionary approach is adopted in the GMBS design. In this respect, a federation of a number of ISPs implementing the necessary functionality can be considered as a first step toward the *QoS-aware mobile Internet*.

1.2.2   Federated ISP Network Architecture

1.2.2.1   Component and network topology

The term *Internet Service Provider* is used to refer to an entity that provides an Internet connectivity service, whether directly to end-users or to other service providers.

The term *Network Service Provider* (NSP) refers to backbone network providers.

Customers are connected to providers via access or hosting facilities in a service provider's Point of Presence (POP).

Different kinds of service providers can be identified, according to geographical constraints:

- *Transit (international) providers*: these connect to other providers in different countries (even in different continents).

- *National providers*: these have POPs serving the different regions of a country.

- *Regional service providers*: these cover specific regions, by connecting themselves to other providers at one or more points.

- *Local service providers*: these provide services directly to end-users.

Transit and national providers belong to the category of NSPs, while regional and local providers can be more correctly considered as ISPs.

The F-ISP network consists of a set of service providers belonging to the four categories described above. Such service providers, suitably interconnected, define peer SLAs to provide mobile QoS sensitive Internet services to a common population of subscribers. The number and the typologies of service providers composing the F-ISP network will depend upon:

- the business arrangements, i.e. settlements, created among them;

- the will of operators to upgrade existing infrastructures to implement the new functionality envisaged in the GMBS for the support of global mobility and E2E QoS.

An evolutionary process is therefore envisaged for the deployment of the F-ISP network where service providers increasingly reach the GMBS community.

Regional service providers (providing connectivity to local service providers) located in remote areas exchange Internet traffic via a service provider backbone network (consisting of transit and/or national service providers).

The backbone network encompasses, as a minimum, the following technical characteristics [HAL-00]:

- *A suitable physical network topology*: the design of the physical topology is such that a consistent, adequate bandwidth is available for all the trajectories among the regional providers exploiting the backbone as trunk carriage for their Internet traffic, even in the event that single or multiple connections become unavailable.

- *Absence of network bottleneck and suitable subscription ratios*: in order to avoid a bottleneck backbone, providers would need to limit over subscription and avoid small tail link circuits towards a POP or a downstream customer; as a common "rule of thumb" a ratio 5:1 between subscriber and available links can be applied (e.g. there should be no more than five T1 links for each T1 backbone connection).

- *Level of network and individual network redundancy*: in order to avoid the situation where a set of destinations become unreachable due to the failure of connections, the network topology should be designed so that each service provider network has POPs connected to multiple Network Access Points (NAPs, see below), other providers' networks and multiple other POPs.

Regional and network service providers (both transit and national) exchange Internet traffic:

- at public NAPs;
- via direct interconnections.

A *NAP* is a high-speed switch or a network of switches to which a number of routers can be connected for the purpose of traffic exchange. The NAP undertakes two main roles:

- *exchange provider* between regional ISPs who want to execute bilateral peering arrangements;
- *transit purchase venue* where regional ISPs can execute purchase agreements with NSPs which, connected to the same NAP, provide backbone connectivity.

Figure 1.2 depicts the interconnection of regional and network service providers at the NAP.

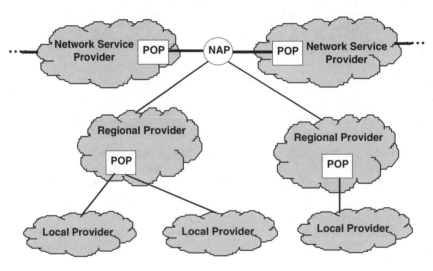

Figure 1.2   Service provider interconnection at NAP.

One basic requirement for the NAP is that it should operate at speeds of at least 100 Mbps. Subsequent upgrading should be possible depending on the overall traffic to be handled. Several technologies could be adopted for the practical implementation of the NAP such as FDDI (Fibre Distributed Data Interface) switches (100 Mbps) or ATM (Asynchronous Transfer Mode) switches (45+ Mbps). The physical configuration of the NAP in the F-ISP network, according to what is currently available in today's Internet, would be a mixture of FDDI, ATM and Ethernet (Ethernet, Fast Ethernet and Gigabit Ethernet) switches. Several different access methods could be adopted such as FDDI, Gigabit Ethernet, DS3 (Data Signal-3 transmission rate 44.736 Mbit/s), ATM OC3 (full duplex data transmission rate up to 155 Mbps) and OC12 (full duplex data transmission rate up to 622.08 Mbps) solutions.

Typically, service providers (both regional and network service providers) manage the routers located in the NAP facilities. Configurations, policies and fees are the responsibility of the *NAP Manager.*

Besides the switching capabilities, the NAP hosts several additional functions and services, including the following.

- *Routing arbiter service.* In order to reduce the requirement for a full peering mesh between all the providers connected to the NAP, a central system named *Route Server* (RS) is co-located at the NAP. In order to set its routing tables, each provider may simply interact with the RS which maintains a database storing the routing information of all the providers connected at the NAP. This eliminates the need of a peering of the providers among each other and, consequently, reduces the exchange of routing information at the NAP. Routing information is exchanged by using the Border Gateway Protocol (BGP), between the RS and the routers which, connected to the NAP, act as BGP speakers of the Autonomous Systems (ASs) involved. An AS is a set of hosts, routers and sub-networks which adopt the same routing policy and which are controlled by the same technical and administrative authority. To the outside world, the entire AS is viewed as a single entity. Each AS has an identifying number that is assigned to it by an Internet Registry [HAL-00]. As shown in Figure 1.3, each BGP speaker exchanges routing information only with the RS and not with all the other BGP speakers at the NAP, thus reducing the burden on processing resources at the routers.

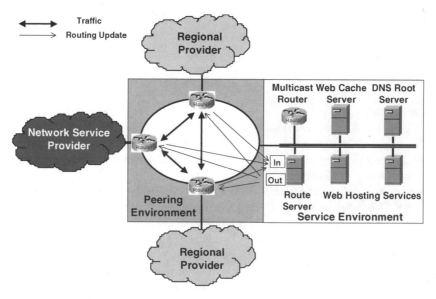

Figure 1.3 Typical NAP infrastructure.

- *Multicast Services.* The NAP may host a *multicast router*. The structure of the NAP makes it particularly suitable for this kind of service.

- *Content Provider Services.* High-volume content providers can be co-located at the NAP. The NAP location guarantees minimum transit distance to a large user population distributed across both the regional service providers directly connected to the NAP and the local providers connected to the regional service providers.

- *Value Added Services*. The NAP environment is particularly suitable for locating Web Cache Server, DNS (Domain Name System) Root Server, Web Hosting Services and so on.

Direct interconnections are realised by establishing direct links between networks, completely avoiding the NAP. Bilateral, peer-to-peer, negotiations usually take place in order to define the locations of these direct interconnections as well as their bandwidth. Direct interconnections provide additional bandwidth between interconnecting networks (with respect to the case where networks interconnect at the NAP), reduce provisioning lead times, increase reliability and considerably scale interconnection capabilities.

1.2.2.2 Edge and core networks

Routers

According to the structure of the overall GMBS, the F-ISP network is considered to be partitioned into an *edge network* and *core network*.

The edge network is the point at which local mobile users are allowed into the network. It consists of a set of local and regional service providers. The core network is a high-speed backbone formed by transit and national service providers. Interconnection between edge and core portions of the F-ISP network takes place:

- at a NAP;

- via direct interconnections.

Routers belonging to the core network are referred to as *core routers* (CRs). The term "core routers" is also used to indicate interior routers of the local and regional service providers forming the edge network, i.e. routers not involved in any direct interconnection between different service providers. These core routers differ from the backbone core routers in the functionality implemented[1].

Routers at the interconnections between edge and core networks are referred to as *border routers* (BRs). Depending on the chosen interconnection typologies, these routers are either NAP routers or routers involved in direct interconnections.

Routers belonging to the edge network involved in a direct interconnection between two regional service providers, between a regional service provider and a local service provider and between two local service providers are referred to as *edge routers* (ERs).

Routers belonging to either a local service provider or a regional service provider and directly interconnecting an end-user are referred to as *leaf routers* (LRs).

IntServ and Diffserv

As far as QoS is concerned, an Integrated Service model based on the RSVP [IET-97a, IET-97b] is adopted at the *edge network*. The RSVP, supported in many commercial

---

[1]For instance, backbone core routers are RSVP capable routers, while regional and local core routers are DiffServ routers.

real-time applications, allows hosts to request quantifiable resources along E2E data paths and provides hard guarantees for the respect of the negotiated QoS level. Since the network must maintain complex per-flow state information in each router, the IntServ approach is viable within small-scale networks. However, at the heart of large-scale networks, such as in high-speed ISP backbones, the cost of state maintenance and of processing and signalling overhead in the routers is overwhelming. For this reason in the F-ISP *core network* a different service model is adopted.

The requirements in terms of speed and scalability of the core network can only be met by implementing a stateless DiffServ model. As DiffServ does not provide strict E2E QoS guarantees, in order to improve the user-perceived QoS performance, an innovative solution called GRIP (Gauge&Gate Reservation with Independent Probing) [IET-01], is implemented in the core portion. GRIP is a fully distributed and scalable admission control scheme, which allows hard QoS guarantees over a stateless DiffServ Internet.

The inter-operation between IntServ and enhanced-DiffServ mechanisms takes place within suitably devised *Gateways* co-located within border routers (at the interconnection between edge and core portion of the F-ISP network). Such an inter-operation requires the implementation of IntServ/DiffServ service mapping and the activation of admission control mechanisms based on the resource availability in the core portion.

The hybrid IntServ/DiffServ solution devised for GMBS is particularly attractive since it:

1. solves scalability problems (DiffServ in the core portion);

2. provides a means to easily adapt to the future evolutions of the Internet network (towards completely DiffServ or IntServ based solutions);

3. allows the configuration of a fully IntServ compatible E2E path in instances where only the edge portion is involved.

Mobility Support
As far as the support of mobility is concerned, Mobile IPv6 [IET-03] is implemented in the edge part of the F-ISP network.

A GMBS user provided with IP-based Terminal Equipment (TE) connects to the Internet network through their own single use GMBS Multi-Mode Terminal (GMMT) or through a collective use GMMT that is simultaneously serving other GMBS users.

The F-ISP network, as an evolution of the current Internet, must offer to users the possibility to use standard Internet applications and content, moreover both QoS and mobility features offer an opportunity for application and content providers to improve their services. In particular, the guaranteed QoS can be used to offer real-time and streaming services of high quality, while mobility can be exploited by providing location-based sensitive services.

As stressed before, it is not possible to guarantee QoS outside the F-ISP. As soon as the requests are directed to the "outside Internet", any QoS requirement is released. For this reason it is very important to implement suitable proxy, caching and mirroring services inside the F-ISP network. In this way, the Internet contents will be accessible without leaving the F-ISP, being guaranteed in terms of QoS.

1.2.2.3   GMBS service node

The GMBS is operated by a GMBS Service Provider (GSP) whose main objective is to provide nomadic GMBS users with connectivity to the Internet network with a guaranteed QoS, irrespective of their position.

The GSP controls a GMBS Service Node (GSN) which is a server belonging to the Internet network of the GSP. It is characterised by an IP address. Communication between the GSN and the Mobile Router (GMMT) takes place by means of IP "signalling" packets transmitted over the wireless link provided by one of the active access segments.

The GSN is in charge of:

- authenticating and registering GMBS users accessing the Internet network via a GMMT. Towards this end, appropriate tables are present in the GSN, containing GMBS user profile information. Such tables are used during the execution of registration and authentication procedures;

- keeping track of the access segment(s) selected by a given GMBS user at any time. The results of the segment selection, reselection and Inter-Segment Handover (ISHO) for each GMBS user are registered in an appropriate table contained in the GSN;

- triggering (optional) the satellite resource release during a satellite to terrestrial ISHO to speed up the release of satellite resources, so avoiding the wait for the expiration of satellite time-out.

- communicating information concerning the resource allocation and congestion status of the access segments to the GMMT. This kind of information is used by the GMMT during the execution of inter-segment mobility procedures such as segment selection, segment reselection and ISHO. The GSN is able to retrieve resource management information from suitably selected access segment nodes (eventually upgraded with additional specific functionality) and elaborate on this. Towards this end, the GSN interfaces with the appropriate satellite, GPRS and UMTS network management nodes. Examples of information retrieved include:

  traffic load on the radio interface (signalling and user traffic);

  usage of resources within the network nodes;

  user activation and use of supplementary services, etc.

The pre-elaboration of these data in the GSN reduces the complexity of the GMMT processing algorithms. In order to retrieve resource management data for the satellite system, the GSN interfaces with the satellite Network Operation Centre (NOC). Communication between the GSN and the NOC takes place by means of IP "signalling" packets transmitted over wired terrestrial links. In order to increase reliability, the Transmission Control Protocol is used as the transport mechanism.

REFERENCES

[CAR-99] F. Carducci, G. Losquadro: The EuroSkyWay Worldwide System providing Broadband Service to Fixed and Mobile End-Users, *International Journal of Satellite Communications*, **17**(2-3), March–June 1999; 143–154.

[HAL-00] S. Halabi, D. McPherson: *Internet Routing Architectures: The definitive BGP resource*, Cisco Press, 2000.

[HUS-99] G. Huston: Interconnection, Peering and Settlements—Part I, *The Internet Protocol Journal*, *Cisco Publications*, **2**(1), March 1999; 2–16.

[IET-01] G. Bianchi, N. Blefari-Melazzi: A Migration Path to Provide End-to-End QoS over Stateless Networks by Means of a Probing-driven Admission Control, *Internet Engineering Task Force*, Internet Draft draft-bianchi_blefari-end-to-end-QoS-02.txt, November 2001.

[IET-03] D.B. Johnson, C. Perkins, J. Arkko: Mobility Support in IPv6, *Internet Engineering Task Force*, Internet Draft draft-ietf-mobileip-ipv6-24.txt, Work in Progress, 30 June 2003.

[IET-97a] R. Braden (Ed.) *et al.*: Resource ReSerVation Protocol (RSVP) – Version 1 – Functional Specification, *Internet Engineering Task Force*, RFC 2205, September 1997.

[IET-97b] J. Wroclawski: The use of RSVP with IETF Integrated Services, *Internet Engineering Task Force*, RFC 2210, September 1997.

ACRONYMS

AS	Autonomous System
ATM	Asynchronous Transfer Mode
BGP	Border Gateway Protocol
BE	Best Effort
BR	Border Router
CR	Core Router
DiffServ	Differentiated Services
DNS	Domain Name System
E2E	End-to-End
ER	Edge Router
F-ISP	Federated - ISP
FDDI	Fibre Distributed Data Interface
GMBS	Global Mobile Broadband System
GMMT	GMBS Multi-Mode Terminal
GPRS	General Packet Radio Service
GSN	GMBS Service Node
GRIP	Gauge&Gate Reservation with Independent Probing
GSP	GMBS SP
IntServ	Integrated Services
IP	Internet Protocol
ISHO	Inter-Segment Handover
ISP	Internet SP
LR	Leaf Router
NAP	Network Access Point
NOC	Network Operation Centre
POP	Point of Presence
QoS	Quality of Service
RS	Route Server
RSVP	Resource Reservation Protocol
SLA	Service Level Agreement
SP	Service Provider
TE	Terminal Equipment
UMTS	Universal Mobile Telecommunications System
W-LAN	Wireless Local Area Network

2

Multi-Segment Access Network

PAOLO CONFORTO, CLEMENTINA TOCCI

Alenia Spazio, Italy

2.1 SATELLITE SEGMENT

2.1.1 Mobile EuroSkyWay

2.1.1.1 *Overview*

The broadband satellite component envisaged in the Global Mobile Broadband System (GMBS) is represented by the mobile component of the EuroSkyWay (ESW) system, also referred to as Mobile EuroSkyWay (M-ESW) [MUR-00]. EuroSkyWay is mainly devoted to fixed users; the mobile component mainly differs from the fixed both in terms of physical layer features, which have been adapted to satisfy the service mobility requirements, and in terms of additional mobility and Quality of Service (QoS) support functionality envisaged for its effective integration in the overall GMBS.

The M-ESW is a satellite-based communications system, forming a continental network supporting a wide range of user data rates, services and applications (see Figure 2.1). Both bursty and continuous bit rate traffic can be supported. Particular care has been devoted to the design of efficient inter-working with current and future terrestrial networks to fully exploit the advantages offered by its unique T-Switch on-board processing (OBP) capability [LOS-98]. This enables the provision of end-to-end (E2E), single hop, meshed connectivity together with on-demand, QoS-guaranteed connections, i.e. the identical connectivity concepts of terrestrial networks have been implemented while still maintaining the advantages of the satellite transmission medium in terms of user data distribution and service deployment.

Moreover, the Dynamic Bandwidth Allocation Control (DBAC) hosted on-board allows the support of a wide range of traffic classes (see the following Section). Since an efficient use of bandwidth is achieved, this directly results in

1. cost saving;

2. high number of contemporary served users;

Space/Terrestrial Mobile Networks.   Edited by R.E. Sheriff, Y.F. Hu, G. Losquadro, P. Conforto, C. Tocci
© 2004 John Wiley & Sons, Ltd   ISBN: 0-470-85031-0

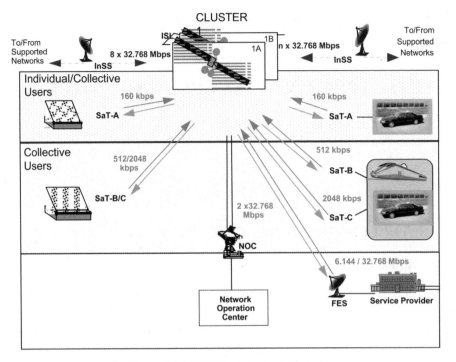

Figure 2.1   M-ESW system architecture.

3. satisfaction within the user population in the usage of the M-ESW communication infrastructure.

M-ESW can operate as both an access and a core network (CN), making it suitable to provide services to end-users and to operate as a backbone interconnecting different telecom operators.

The main aspect of the M-ESW system to be highlighted in this context, especially with regard to the impact on the design of an integrated Satellite/Cellular/Internet scenario, is that M-ESW acts as an *underlying* network with respect to the existing terrestrial networks it is connected to. M-ESW provides the means to transparently transfer both signalling and traffic data generated both at terminal level and in external networks which, for this reason, can be logically considered as *overlaying* networks (OLN). The inter-operation with the terrestrial OLN is realised by means of *Network Interface Units* (NIU), which are able to accept not only any existing protocols but also any possible new protocols. The second fundamental aspect to be noted is that the M-ESW network provides the above mentioned transparent connectivity by operating through a minimum set of protocols and signalling procedures which are applicable to the satellite system. This enables efficient control of the provision of switched capacity on demand and for mobility/localisation purposes. Resources are allocated through an efficient bandwidth-on-demand mechanism based on the transmission to the *Traffic Resource Manager* located within the payload of appropriate *In Band Requests* guaranteeing low latency of the assignment process. From the observations above, it can be seen that the M-ESW network architecture can be used to provide an implementation of the 'Internet-in-the-sky' concept. In the following, the major features of this architecture are briefly described.

The M-ESW network is organised into physical sub-networks (M-ESW-SNs), an M-ESW-SN being served by a single satellite. Each M-ESW-SN interacts with

1. one or more external networks, e.g. Internet, to extend the served user community to include also those users connected to the terrestrial communication infrastructure and,

2. other M-ESW-SNs to broaden the user community to include M-ESW users which are served by a different satellite.

The complete set of communication capabilities allows the M-ESW system to connect a user of an M-ESW-SN both to other end-users and to service providers. This is regardless of whether the end-users/service providers are directly connected to the same M-ESW-SN, or to another M-ESW-SN or to an external (terrestrial/satellite) network. The M-ESW network provides mobility services to whatever terrestrial protocol it supports, i.e. any protocol a user application is using. This includes assigning IPv6 addresses to M-ESW users/terminals. It provides the user terminal mobility completely transparent to the terrestrial protocol's switching point that the user is connected to, i.e. no customisation of the terrestrial protocols is needed.

A specific set of terminals is envisaged to cope with the interface requirements, in particular with an appropriate transmit/receive data rate that the interconnection with terrestrial networks imposes on the system.

2.1.1.2 Network nodes

In the following, the network nodes of the M-ESW system are listed, specifying some of the main functionality implemented:

The *Network Operation Centre* (NOC) performs satellite connection handling by implementing, as a centralised entity, the execution of the following procedures:

- *Synchronisation Procedures*: the NOC assigns the channels between the user terminals and satellite devoted to the acquisition and maintenance of synchronisation.

- *Registration and Authentication Procedures*: the NOC performs both terminal equipment identification and user authentication.

- *Connection Management Procedures*: these consist of connection control procedures, i.e. call admission control (CAC) (executed to verify if the system is able to provide the necessary resources to support the connection with the required QoS), connection set-up and connection clearing and resource management procedures (based on the implementation of a bandwidth-on-demand mechanism).

The NOC has full visibility of the network configuration and the connections, both active and those being set-up. Information supporting synchronisation, registration and connection management must be available. Thus, the NOC is equipped with dedicated databases, in order to store appropriate system information, supporting all procedures described above. These databases contain the following tables:

- *Synchronisation Table*: stores the list of channels assigned to each terminal for synchronisation maintenance;

- *Localisation Table*: stores information regarding the user location within the satellite coverage area;

- *Registration and Authentication Tables*: contains the list of legal terminals and the authentication keys assigned, at subscription time, to the M-ESW users;

- *Connection Assignment Table*: stores information supporting the unequivocal identification of satellite connection, both "active" and "in progress".

The *Fixed Earth Stations* (FESs) represent the point of attachment to the Internet network and therefore, they implement inter-working functionality between the satellite and terrestrial protocols. This is realised by means of programmable, and therefore reconfigurable and updateable, NIUs. The NIU is also in charge of performing synchronisation, location management and connection management procedures. Since the M-ESW protocols are decoupled from those of the terrestrial network, it is sufficient to re-program the inter-working functionality contained in the NIU to allow inter-operation.

The *OBP-based Satellite Payload*, in addition to switching functionality, is also in charge of executing synchronisation and resource management procedures.

The *Broadband Satellite Terminals* (SaTs) are classified into three types based on their maximum sustained information rate and usage:

- Type-A (both for individual and collective use) with uplink bit rate of 160 kbps and downlink bit rate of 6 Mbps;

- Type-B (for collective use) with uplink bit rate of 512 kbps and downlink bit rate of 16 Mbps;

- Type-C (for collective use) with uplink bit rate of 2 Mbps and downlink bit rate of 16 Mbps.

The M-ESW segment is interconnected to the Federated Internet Service Provider (F-ISP) network. M-ESW is seen, from an Internet perspective, as a normal service provider since it is able to provide Internet connectivity. According to the terminology introduced above, M-ESW is, at the same time, an ISP, since it is able to provide connectivity directly to end-users, and a Network Service Provider (NSP), since it can act as a backbone network provider. Therefore, M-ESW connects to the F-ISP network in the same way as each service provider belonging to this network does, that is (see Figure 2.2):

- with direct interconnection with terrestrial service providers: in this case direct links are established between a (or more) M-ESW FES(s) and a (or more) edge router(s) of the terrestrial service provider. Service Level Agreements (SLAs) are negotiated bilaterally on a peer-to-peer basis between M-ESW and the terrestrial service provider. The objective of this negotiation is to obtain (each others) guarantees about a given level of performance and availability. This agreement is realised by fixing the value of all the

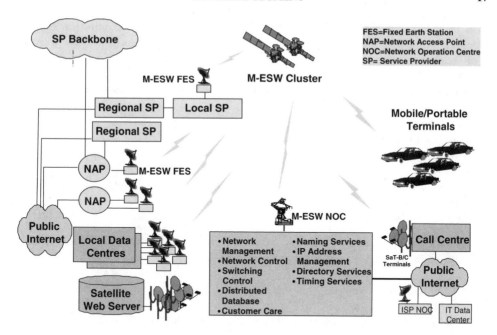

Figure 2.2 M-ESW interconnection to a federated ISP network.

parameters associated with the additional functionality implemented in the GMBS for the support of provided mobile QoS-sensitive Internet services;

- at the Network Access Point (NAP): in this case the M-ESW FES connects to a NAP which represents a neutral meeting ground where service provider traffic is exchanged. M-ESW can exploit this meeting ground in order to execute purchase agreements with other service providers connected at the same NAP. In this context, M-ESW typically acts as a NSP, providing backbone connectivity to the regional service providers at the NAP.

Direct interconnections provide additional bandwidth between interconnecting networks (with respect to the case where networks interconnect at the NAP), reduce provisioning lead times, increase reliability and considerably scale interconnection capabilities. A drawback is related to the costs associated with the installation and maintenance of the large amount of infrastructure necessary to implement direct interconnections, which are realised to serve a single service provider.

One of the benefits of NAP connections is that the same infrastructure, i.e. the same M-ESW FES can be used to exchange traffic with several service providers (those connected at the NAP). Moreover, once the connection at the NAP has been realised, purchase agreements can be realised in a time effective manner and without any additional costs for further infrastructure deployments.

It is worth noting that the M-ESW network architecture can be integrated with the existing terrestrial communication infrastructures due to its advanced on-board switching scheme and the flexible inter-working functions at the terminal side. This appears particularly advantageous for the Internet Services Architecture, where the current ISP network infrastructure will be easily complemented by the satellite-based communication infrastructure. The satellite component will upgrade and complement the terrestrial infrastructure with E2E QoS-guaranteed connectivity services without the need of any customisation, other than the insertion of a SaT at suitably selected nodes.

From an Internet topological viewpoint, the whole M-ESW network acts as an Internet *Autonomous System* (M-ESW-AS).

Since M-ESW acts both as an ISP providing connectivity directly to end-users, and as a NSP, i.e. as a backbone network provider, it is configured as *Multi-Homed Transit AS*.

This means that the M-ESW AS:

- can be characterised by several connections with the outside world, one connection for each M-ESW FES connected to a NAP or, via a direct interconnection, to an edge router of a service provider;

- can allow transit traffic by other ASs (transit traffic is any traffic that has an origin or a destination which does not belong to the M-ESW AS).

M-ESW exchanges routing information with other ASs using the Border Gateway Protocol (BGP).

Each M-ESW FES, representing the connection with a different (terrestrial) AS, acts as a BGP speaker.

Basically three different AS typologies can be identified, depending on the network configurations the AS refers to:

- *Stub or Single-Homed AS*: this reaches networks outside its domain via a single exit point;

- *Multi-Homed Nontransit AS*: this has more than one exit point to the outside world and does not carry transit traffic;

- *Multi-Homed Transit AS*: this has more than one exit point to the outside world and carries transit traffic.

M-ESW support to the Internet architecture results in relevant performance improvements due to advanced technical solutions such as

1. intelligent routing, i.e. the optimisation of the IP flow landing point in order to minimise the terrestrial path between the satellite and terrestrial end-points of the IP flow;

2. ad hoc QoS support based on the advanced M-ESW traffic management capabilities to maximise the performance of the appropriate mechanisms of the Internet.

Both of these solutions will strongly mitigate, if not eliminate, the impact of the terrestrial paths on the Internet E2E QoS.

The M-ESW traffic management is based upon a distributed and integrated architecture:

1. it is "distributed" in the sense that it is composed of a set of functions which are hosted by different elements of the M-ESW network;

2. it is "integrated" by being embedded into the different system functions which manage the different phases of the M-ESW connection life cycle.

The DBAC plays a key role in the overall management of the traffic resource. It is a priority-based scheme based on a bandwidth allocation on request and allowing a resource utilisation close to 100% for providing QoS-guaranteed connections to High Priority (HP) traffic sources and redistributing the unused traffic resources to Low Priority (LP), best effort (BE) traffic sources. This results in a very high utilisation of the satellite traffic resources together with no loss phenomena due to congestion. This is due to the capacity assigned to the HP traffic being simply "lent" to the LP sources, i.e. every time the HP source needs the full capacity, the loaned traffic resources will be pre-empted to the BE owner and given back to the HP source.

The traffic management provides the following functions, which co-operate to obtain a more effective resource utilisation (see Figure 2.3):

- *Connection Admission Control*: this is defined as the set of actions taken by the network during the connection set-up phase (or during connection re-negotiation phase) in order to determine whether a satellite connection request can be accepted or rejected (or whether a request for re-allocation can be accommodated).

- *Usage Parameter Control (UPC)*: this is defined as the set of actions taken by the network to monitor and control traffic, in terms of traffic offered and validity of the satellite connection, at the end-system access. Its main purpose is to protect network resources from malicious as well as unintentional misbehaviour, which can affect the QoS of other already established connections, by detecting violations of negotiated parameters and taking appropriate actions.

- *Traffic Resource Management (TRM)*: the satellite system provides traffic resource on-demand. The amount of allocated resource depends on: (i) the requested amount of resources, (ii) the connection traffic contract, (iii) the user profile of the owner of the connection.

- *Feedback Controls*: the satellite system provides a set of actions taken by the network and by the end-systems to regulate the traffic submitted on satellite connections according to the state of network elements.

The TRM dynamically assigns bandwidth with a granularity of 16 kbps for user terminals and 64 kbps for provider terminals. This granularity allows the provision of a very low bit rate appropriate to the Internet return path. The adoption of TDMA (Time Division Multiple Access) schemes based on frame multiples allows further scaled down granularities.

2.1.1.3   Satellite access technique, modulation and coding

The satellite access technique is Multi-Frequency TDMA (uplink) and Time Division Multiplex (downlink).

Quadrature Phase Shift Keying (QPSK) modulation is used on both uplink and downlink, since it provides good performances in terms of

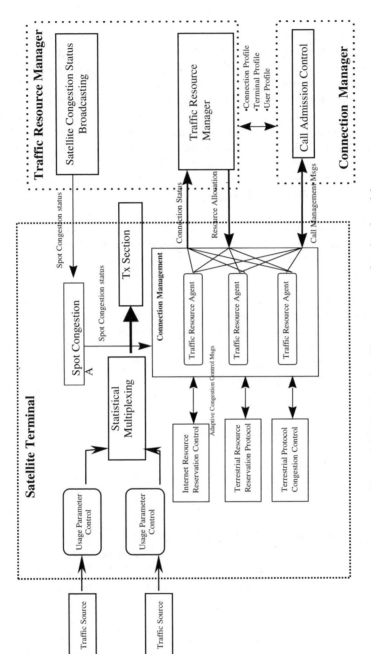

Figure 2.3 Distributed traffic management functional model.

- spectral efficiency;

- robustness against synchronisation errors due to time jitter, phase noise and so on;

- low implementation complexity.

As far as coding is concerned, Forward Error Correction coding is adopted: Reed-Solomon (RS) [76, 60] + parity check 10/9 coding on the uplink and Digital Video Broadcasting-Satellite coding on the downlink is employed.

2.1.1.4 User plane

Figure 2.4 shows: (a) the M-ESW network elements involved in the traffic data exchange between end-users; and (b) the M-ESW user plane protocol stack.

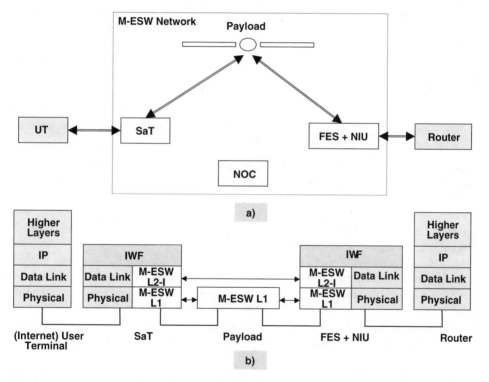

Figure 2.4 (a) Network elements involved in the traffic data exchange; (b) M-ESW user plane protocol stack.

The data exchange within the M-ESW network is supported by *the M-ESW Layer 2-I* representing the particular instance of the whole M-ESW Layer 2 dedicated to the data exchange between satellite terminals (both FESs and SaTs). M-ESW Layer 2-I entities perform formatting/de-formatting, error recovery and flow control functions. The M-ESW Layer 2-I can provide two forms of data transfer service, depending on the constraints coming from the higher layer: acknowledged or unacknowledged.

The *IWF layer* hosts the entities in charge of the inter-working between the higher layer protocol, such as IP, and the M-ESW specific protocol.

The M-ESW network elements involved in the support of the connection in the M-ESW portion are the SaT, the payload and the FES. Formatting/de-formatting, error recovery and flow control functions are only executed at the M-ESW connection end-points, i.e. in the SaT and in the FES.

2.1.1.5   Control plane

The M-ESW system has been devised to provide efficient connection-oriented satellite communication services to OLN applications or native M-ESW applications. The connection-oriented feature implies that in order to support data exchange between M-ESW terminals (both FESs and SaTs) a set-up phase aimed at establishing a logical satellite connection has to be undertaken. Then during the connection lifetime, appropriate connection maintenance procedures aimed at dynamically allocating resources and at renegotiating connection parameters have to be executed. At the end of data transfer, the connection is torn down, freeing the resources through a termination procedure.

The above-mentioned procedures belong to the class of *Connection Management* (CM). The CM functionality is further divided into two separate sub-functions:

- *Connection Control (CC)*: pertaining to CAC, connection set-up and connection termination. The entities in charge of executing these functions reside in the terminals, in the NOC and in the TRM within the satellite payload;

- *Resource Management (RM)*: pertaining to the dynamic assignment and release of resources required for a connection once it has been established. The corresponding entities reside in the terminal and in the TRM.

Figure 2.5 depicts: (a) the protocol stacks for the CC and (b) the RM procedures. CC and RM entities belong to the M-ESW Layer 3.

At terminal switch-on, a user must log onto the M-ESW network by activating *Registration and Authentication procedures*. The successful completion of these procedures allows the M-ESW network to locate active terminals throughout the coverage area and at the same time prevent illegal users from gaining access to the network resources.

The entities in charge of executing registration and authentication procedures reside in the terminals and in the NOC. As can be observed in Figure 2.5(c), such entities belong to the M-ESW Layer 3.

It is worth highlighting that the M-ESW signalling procedures previously described are completely managed within the M-ESW network. This implies that the OLN, i.e. the Internet Network, is not involved.

2.1.2   Low Earth Orbit Extension

There may be occasions within the GMBS coverage area, where the combination of M-ESW and terrestrial mobile networks is not able to provide services to the user. Terrestrial networks, such as GPRS and UMTS, will gradually be deployed, focusing initially on areas

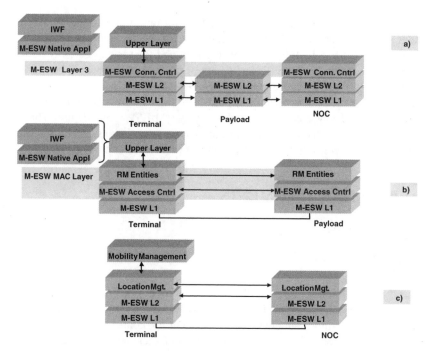

Figure 2.5 (a) M-ESW connection control procedure; (b) M-ESW resource management procedure; (c) M-ESW registration/authentication procedure protocol stacks.

of high usage. Similarly, M-ESW component is not able to guarantee complete global coverage to all regions of the world, especially in the polar regions.

To meet the need for global coverage, a hybrid satellite constellation concept could be utilised, comprising of a Low Earth Orbit (LEO) satellite component connected to the M-ESW satellite component via an inter-orbit link. Connection to the core network would be achieved through normal gateways.

A possible hybrid system architecture is depicted in Figure 2.6. The LEO segment provides an access point to the network for those users located in areas where the M-ESW

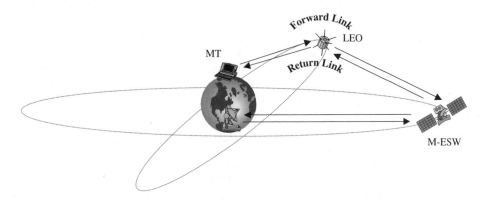

Figure 2.6 Possible hybrid constellation basic architecture.

satellite is not able to provide a service. In the forward link, the data received by the LEO component from one or more Mobile Terminals (MTs) will reach the final destination through an inter-orbit link with the M-ESW segment and the serving gateway. In the return link, the data received by the M-ESW component from the gateway would be transmitted to the final user through the LEO component.

The main goal of such an architecture would be to increase total coverage by adding just one, or a few, small, low cost LEO satellites to the M-ESW architecture. The coverage offered by the LEO component would be discontinuous, requiring a store-and-forward mode of operation, if a full constellation of satellites were not to be deployed.

2.2   TERRESTRIAL SEGMENTS

2.2.1   GPRS Segment

2.2.1.1   Overview

The General Packet Radio Service (GPRS) system provides complementary coverage with respect to that of the satellite. A GMBS user, provided with a GMBS Multi-Mode Terminal (GMMT), selects the most suitable access segment to connect to the F-ISP network. Selection is executed on the basis of several factors (depending on the particular user profile) such as environment (and therefore link availability), type of service and cost effectiveness.

GPRS is a relatively new bearer service allowing efficient wireless access to packet data networks (PDNs), e.g. Internet. It uses a packet mode technique to efficiently transfer high speed and low speed data and signalling between mobile stations (MSs) and external data networks.

GPRS guarantees a flexible allocation of radio resources. It allows the subscriber to send and receive data in an E2E packet transfer mode, without using any network resources in circuit-switched mode. This allows for autonomous operation of GPRS and best fits the bursty traffic characteristics.

GPRS has evolved from the European second-generation GSM (Global System for Mobile communications) network and, due to the adoption of packet radio principles, it improves the utilisation of radio resources, offers volume based billing, higher transfer rates, shorter access time and simplifies the access to packet data [BET-99, SAL-99].

New radio channels, called Packet Data Traffic Channels (PDTCH), have been defined for GPRS, which are mapped onto the same physical channels as GSM. Nevertheless, the allocation of these channels is flexible since from one to eight radio interface time-slots can be allocated per TDMA frame. The active users share time-slots [ETS-02b]. This implies that bit rates ranging from 9 kbps up to 150 kbps can be granted to the same user. The radio interface resources are dynamically assigned, a channel being allocated only when needed and immediately released after the transmission of packets.

With respect to the GSM architecture, GPRS (see Figure 2.7) introduces two new network nodes which are responsible for the delivery and routing of data packets (i.e. IP packets in the GMBS) between GPRS MSs and the external PDNs (i.e. the F-ISP network). In particular, the *Serving GPRS Support Node* (SGSN) is responsible for the delivery of data packets from and to the MSs within its service area. The location register of the SGSN stores

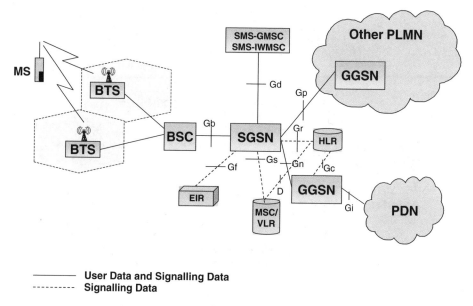

Figure 2.7 GPRS system architecture.

location information (e.g. current cell, current Visitor Location Register (VLR)) and user profiles (e.g. International Mobile Subscriber Identity (IMSI), address(es) used in the PDN) of all GPRS users registered with this SGSN. The *Gateway GPRS Support Node* (GGSN) represents rather the gateway towards the F-ISP network and therefore it executes all the functions necessary for inter-working. It converts the GPRS packets coming from the SGSN into the IP format and sends them out on the corresponding PDN. In the reverse direction, Packet Data Protocol (PDP) addresses of incoming data packets are converted to the GSM address of the destination user. The re-addressed packets are sent to the responsible SGSN. For this purpose, the GGSN stores the current SGSN address of the user and his/her profile in its location register. The GGSN also performs authentication and charging functions.

Each SGSN can be connected to many GGSNs and *vice-versa*. In the former case, an SGSN may route its packets over different GGSNs to reach the F-ISP network while, in the latter case, the same GGSN acts as an interface toward the F-ISP network for several SGSNs.

All the GPRS Support Nodes (GSNs), both SGSN and GGSN, are connected via IP-based GPRS backbone networks. Two kinds of GPRS backbone are envisaged (see Figure 2.8):

- *Intra-PLMN (Public Land Mobile Network) backbone network*: connecting GSNs belonging to the same PLMN; it is therefore a private IP-based network of the GPRS network providers;

- *Inter-PLMN backbone network*: connecting GSNs belonging to different PLMNs; in this case a roaming agreement between GPRS network providers is necessary to install the backbone.

The gateways between the PLMNs and the external inter-PLMN backbone are called *border gateways*.

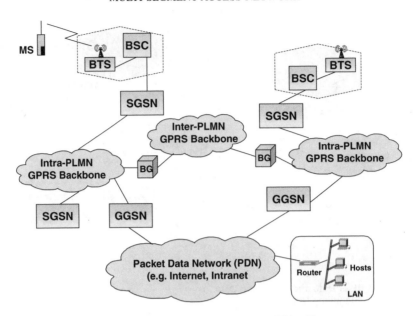

Figure 2.8   GPRS intra- and inter-PLMN backbone.

Two different encapsulation schemes are used by GPRS. Firstly, all packets are encapsulated between the GSNs by means of a GPRS Tunnelling Protocol (GTP). This is to enable usage of different PDPs, even if these protocols are not supported by all SGSNs. Secondly, encapsulation between an MS and SGSN is performed to decouple the logical link management from the network layer protocols.

As shown in Figure 2.7, the SGSN is connected to the *Base Station Subsystem*, which presents the same structure as envisaged in the GSM network.

As far as the interfaces between network nodes are concerned (see Figure 2.7):

- the *Gb-interface* connects a Base Station Controller (BSC) with the SGSN;

- the *Gn-interface* is used to exchange user and signalling data between GSNs belonging to the same PLMN; in particular, this interface allows the SGSNs to exchange user profiles when an MS moves from one SGSN area to another;

- the *Gp-interface* is used to exchange user and signalling data between GSNs belonging to different PLMNs;

- the *Gf-interface* connects the SGSN to the Equipment Identity Register (EIR); the SGSN may query the IMEI (International Mobile Equipment Identity) of an MS trying to register with the network;

- the *Gi-interface* connects the PLMN with external public or private PDNs, (i.e. with the F-ISP network in the GMBS). Even though both interfaces to IPv4 and IPv6 are supported, IPv6 is assumed as baseline in the target GMBS;

- the *Gr-interface* is used to exchange information between the Home Location Register (HLR) and SGSN (for example, the SGSN informs the HLR about the current location of an MS; when the MS registers with a new SGSN, the HLR will send the user profile to the new SGSN);

- the *Gc-interface* provides a signalling path between GGSN and HLR (for instance, the GGSN queries a user's location and profile information in order to update its location register).

Other interfaces are envisaged in order to have efficient coordination between packet-switched (GPRS) and circuit-switched (conventional GSM) services:

- the *Gs-interface* connects the databases of the SGSN and Mobile Switching Centre (MSC)/VLR;

- the *Gd-interface* connects the SGSN with the SMS Gateway MSC (SMS-GMSC).

2.2.1.2 Access nodes

In the following, the GPRS network nodes are listed, specifying some of the main implemented functionality [ETS-02b]:

- *Serving GPRS Support Node (SGSN)*: this is responsible for the delivery of data packets from and to the MSs within its service area. The SGSN executes network access control functions (i.e. authentication and authorisation, admission control, charging data collection), packet routing and transfer functions (i.e. relay, routing, address translation and mapping, encapsulation, tunnelling, compression and ciphering), mobility management functions, logical link management functions (logical link establishment, logical link maintenance and logical link release) and radio resource management functions (i.e. path management).

- *Gateway GPRS Support Node (GGSN)*: this represents the gateway towards the F-ISP network. The GGSN executes network access control functions (i.e message screening and charging data collection), packet routing and transfer functions (i.e. relay, routing, address translation and mapping, encapsulation, tunnelling) and mobility management functions. In particular, the GGSN is able to assign IP addresses to the GPRS MSs.

- *Base Station SubSystem*: this consists of Base Transceiver Stations (one for each cell) and a BSC. The Base Station SubSystem executes network access control functions (i.e. admission control), packet routing and transfer functions (i.e. relay and routing) and radio resource management functions.

- *Home Location Register*: this is a database used to store the user profile, the current SGSN address and the PDP address(es) for each GPRS user in the PLMN. The HLR is involved in the execution of network access control functions (i.e. registration, authentication and authorisation), packet routing transfer functions (i.e. ciphering) and mobility management functions.

- *Mobile Station*: this consists of two main blocks: (i) the MT providing the access to the GPRS radio resources; and (ii) the Terminal Equipment (TE) hosting IP and upper layer functionality. The MS (in particular the MT) executes network access control functions (i.e. authentication and authorisation, admission control and packet terminal adaptation), packet routing and transfer functions (i.e. relay, routing, address translation and mapping, encapsulation, compression and ciphering), mobility management functions, logical link management functions (i.e. logical link establishment, logical link maintenance and logical link release) and radio resource management functions.

- *Border Gateway (BG)*: this connects a PLMN to the external inter-PLMN backbone. The BG provides the necessary inter-working and routing protocols (e.g. BGP). Moreover, the BG performs security functions to protect the private intra-PLMN backbones against unauthorised users and attacks.

2.2.1.3   GPRS QoS support

The QoS requirements of typical mobile packet data applications are very diverse (e.g. consider real-time multimedia, Web browsing and e-mail transfer). Support of different QoS classes, which can be specified for each individual session, is therefore an important feature. GPRS allows the definition of QoS profiles using the following data transfer attributes:

- *Service Precedence*: can be high, normal or low; this is used to determine which packets will be discarded under congestion conditions.

- *Reliability*: indicates the transmission characteristics that are required by an application. Three reliability classes are defined in terms of the probability of losing packets, or obtaining duplicate, out of sequence or corrupt packets. Table 2.1 lists the three classes of data reliability. An application with good error correction capabilities might use class three, while an error-sensitive application with no error correction might use class one.

- *Delay*: defines the E2E transfer delay incurred in the transmission of Service Data Units (SDUs) through the GPRS network. This includes the radio channel access delay (on uplink) or radio channel scheduling delay (on downlink), the radio channel transit delay (uplink and/or downlink paths) and the GPRS network transit delay (multiple hops). It does not include transfer delays in external networks (Internet). Four delay classes are defined, three predictive and one BE. The predictive classes define different mean and 95

Table 2.1   GPRS reliability classes

Reliability class	Lost SDU probability	Duplicate SDU probability	Out of Sequence SDU probability	Corrupt SDU probability	Example of application characteristics
1	10^{-9}	10^{-9}	10^{-9}	10^{-9}	Error sensitive, no error correction capability, limited error tolerance capability.
2	10^{-4}	10^{-5}	10^{-5}	10^{-6}	Error sensitive, limited error correction capability, good error tolerance capability.
3	10^{-2}	10^{-5}	10^{-5}	10^{-2}	Not error sensitive, error correction capability and/or very good error tolerance capability.

Table 2.2   GPRS delay classes

Delay Class	Delay (maximum values)			
	SDU size: 128 octets		SDU size: 1024 octets	
	Mean Transfer Delay (s)	95 percentile Delay (s)	Mean Transfer Delay (s)	95 percentile Delay (s)
1. (Predictive)	< 0.5	< 1.5	< 2	< 7
2. (Predictive)	< 5	< 25	< 15	< 75
3. (Predictive)	< 50	< 250	< 75	< 375
4. (Best Effort)	Unspecified			

percentile delays for small (128 octet) and large (1024 octet) packets, while delay is unspecified for the BE class; the relevant delay values are reported in Table 2.2.

- *Throughput*: defined in terms of maximum bit rate or mean bit rate (the latter includes the idle periods in 'bursty' traffic) and is requested by the user. The two values can be negotiated (and renegotiated through a session) up to the Information Transfer Rate (which itself ranges from one to eight traffic channels, as noted earlier). The maximum transfer rate using 14.4 kbit/s traffic channels is in the region of 150 kbit/s in total, but this might be limited commercially to around half this value.

2.2.1.4   GPRS access technique, modulation and coding

As in GSM, GPRS use a combination of Frequency Division Multiple Access and TDMA. Gaussian Minimal Shift Keying modulation is employed.

Four different coding schemes (CS) are defined to protect the transmitted data packets against errors: CS-1, CS-2, CS-3 and CS-4.

It is worth highlighting that the mapping of user application QoS parameters onto GPRS attributes is an implementation issue (that is not defined in GPRS) and defaults will be used if parameters are not defined.

Moreover, GPRS is able to respond to local data traffic conditions adaptively. In particular, GPRS includes the functionality to increase or decrease the amount of radio resources allocated to it on a dynamic basis. Nevertheless, the criteria used to decide on dynamic changes of the GPRS part of the radio resource and the necessary procedures, including radio protocol and timers, needed to perform the change of radio resources, has not been specified in GPRS. The only requirement is that within GPRS the dynamic allocation of the radio resource for bursty or lengthy file transfer applications shall be such that it can be controlled by the network operator.

The following traffic management functions are envisaged in the GPRS system:

- *Call Admission Control*: this is based on available resources;

- *Resource Allocation*: dynamic modification of a PDP context/radio bearer reconfiguration;

- *Scheduling*: this is implementation dependent.

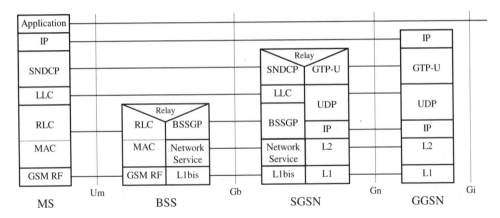

Figure 2.9   GPRS user plane protocol stack.

2.2.1.5   User plane

Figure 2.9 depicts the user plane protocol stack for the GPRS system. The Gi-interface represents the interface towards the Internet (edge) router that the GGSN is connected to. The *GTP Layer* hosts the entities in charge of tunnelling user data and protocols between GSNs in the GPRS backbone. The IP packets coming from the edge router are encapsulated in GTP PDUs and delivered to the *TCP/UDP Layers*, which are in charge of transporting them over the GPRS backbone network.

The lower IP layer represents the GPRS network protocol used to route user data and control signalling.

2.2.1.6   Control plane

Figure 2.10 shows the control plane protocol stack for the GPRS network. The *GPRS Mobility Management and Session Management* (GMM/SM) hosts the entities in charge of

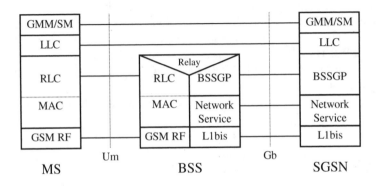

Figure 2.10   GPRS control plane protocol stack.

supporting mobility management functionality, such as GPRS attach/detach, security, routing area update, location update and PDP context activation/de-activation.

Like the M-ESW system, the GPRS signalling procedures are completely managed by the GPRS specific entities, i.e. the Internet Network is not involved.

2.2.2 UMTS Segment

2.2.2.1 Overview

The Universal Mobile Telecommunications System (UMTS) represents the European third-generation solution for the wireless terrestrial access segment.

The rationale of this scenario is that, as with GPRS, UMTS provides complementary coverage with respect to that of the satellite. As before, a GMBS user, provided with a GMMT, selects the most suitable access segment, i.e. M-ESW or UMTS, to connect to the F-ISP network. Selection is executed based on the criteria previously described under GPRS.

The UMTS infrastructure is divided into the *Access Network Domain*, which is characterised by being in direct contact with the *User Equipment* (UE), and the *Core Network* (CN) *Domain* [ETS-01]. This division is intended to simplify the process of de-coupling access related functionality from non-access related functionality.

The CN Domain is divided, from a functional point of view, into a *Packet Switched* (PS) *Service Domain* and a *Circuit Switched Service Domain*. Only the PS Service Domain is considered in the framework of GMBS.

The overall network is shown in Figure 2.11[ETS-03b].

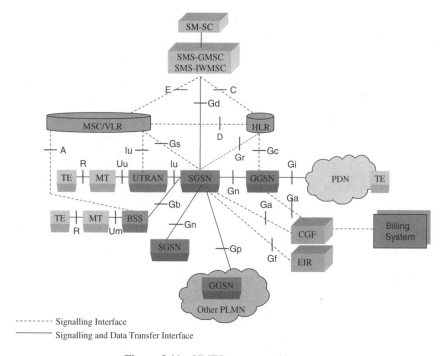

Figure 2.11 UMTS system architecture.

The PS service domain only transfers PS data offering PS type of connection and uses the PS domain specific entities, i.e. *SGSN* and the *GGSN*. These entities provide the PS domain CN functionality. In particular:

- *3G Serving GPRS Support Node*: this is the serving node that supports both the Base Station SubSystem and the Radio Network Subsystem (RNS) access network types;

- *3G Gateway GPRS Support Node*: this is a gateway to an external PDN.

Moreover, the UMTS CN is provided with several nodes, which have database functionality:

1. the *Home Location Register*, which stores subscription and location information;

2. the *Authentication Centre* which is associated with a HLR and generates the valid data for subscriber authentication, and;

3. the *Equipment Identity Register* (EIR), which stores the IMEI used in the UMTS and ensures the pursuit and the blocking of stolen and defunct MSs.

Two different types of access network are used: the *RNS* and, for backward compatibility with second-generation systems, the *Base Station SubSystem*. The *UMTS Terrestrial Radio Access Network* (UTRAN) is used in UMTS. The UTRAN architecture is shown in Figure 2.12.

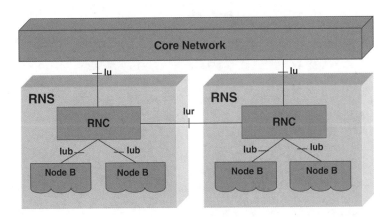

Figure 2.12   UTRAN architecture.

The RNS consists of one or more Node Bs and one Radio Network Controller (RNC). The RNC is the network entity of a PLMN with the functionality for control and access of one or more Node Bs. A Node B is the network entity that serves one or more cells and connects to the MS. A Node B can support Frequency Division Duplex (FDD) mode using Wideband Code Division Multiple Access (W-CDMA) and Time Division Duplex (TDD) mode using Time Division CDMA.

2.2.2.2   Access nodes

In the following, the network nodes of the UMTS are listed, specifying some of the main functionality implemented:

- *3G Serving GPRS Support Node (SGSN)*: this is responsible for the delivery of data packets to and from the MSs within its service area. Subscription information (e.g. IMSI, subscribed PDP address), location information (e.g. GGSN address, VLR number) and session information (e.g. used PDP address, QoS profile) are stored in the SGSN, which is necessary to handle the originating and terminating packet data transfer. The SGSN executes network access control functions (i.e. authentication and authorisation, admission control, charging data collection), packet routing and transfer functions (i.e. relay, routing, address translation and mapping, encapsulation and tunnelling) and mobility management functions.

- *3G Gateway GPRS Support Node (GGSN)*: this represents the gateway towards the F-ISP network. It stores data necessary to route data packets to and from the external PDN. This data includes subscriber information (e.g. IMSI, subscribed PDP address), location information (e.g. SGSN address) and session information (e.g. used PDP address, QoS profile, packet filter). Upon reception of a data packet from the PDN, the GGSN evaluates the PDP address and sends the packet to the corresponding SGSN. The GGSN executes network access control functions (i.e message screening and charging data collection), packet routing and transfer functions (i.e. relay, routing, address translation and mapping, encapsulation, tunnelling) and mobility management functions. In particular the GGSN is able to assign IP addresses to the UMTS MSs.

- *UMTS Terrestrial Radio Access Network* [ETS-02a]: This consists of a set of RNSs and is connected to the CN via the Iu-inferface. The RNS executes network access control functions (i.e. admission control), packet routing and transfer functions (i.e. relay, routing, address translation and mapping, encapsulation and tunnelling) and radio resource management functions.

- *Home Location Register*: this is a database used to store the user profile, the current SGSN address, and the PDP address(es) for each UMTS user in the PLMN. The HLR is involved in the execution of network access control functions (i.e. registration, authentication and authorisation), packet routing transfer functions (i.e. ciphering) and mobility management functions.

- *Mobile Station*: this consists of the Mobile Equipment, which, in turn, is composed by MT, Terminal Adapters, TE, and the Subscriber Identity Module. The MS executes network access control functions (i.e. authentication and authorisation, admission control and packet terminal adaptation), packet routing and transfer functions (i.e. relay, routing, address translation and mapping, encapsulation, compression and ciphering), mobility management functions and radio resource management functions.

2.2.2.3   UMTS QoS support

The QoS approach adopted in the UMTS is based on an E2E paradigm implemented by a QoS–related architecture composed of several levels.

E2E services are characterised by a certain QoS profile requested by the user. In order to support such a QoS profile, a Bearer Service has to be selected. Ad-hoc characteristics and functionality, from the sender to the receiver of the service define this radio bearer.

The UMTS QoS architecture is based on an Integrated Service approach envisaging the execution of a CAC operation for each connection, which is to be set-up by the network in order to obtain the necessary resources to support the application [ETS-03e]. This connection is set-up by means of a *PDP Context Activation procedure*. During the setting of the PDP context, several lower layer operations are executed, such as the creation of a radio bearer, which is able to sustain the QoS constraints of the application requesting the service. Therefore, a PDP context activation procedure may fail when it is not possible to allocate the appropriate amount of requested resources for the new radio bearer.

Within the QoS profile, the specification of the traffic attributes, which define the type of radio bearer that is necessary to sustain the application, are carried on the PDP context activation request message. The traffic attributes are as follows:

- *Traffic class* ('Conversational' and 'Streaming' for conversational real-time services, 'Interactive' and 'background' for WWW, e-mail, Telnet, FTP, News and so on): traffic Class is itself a service attribute since it can be used by UTRAN to make assumptions about traffic features of the traffic source and to optimise transport, based on those features;

- *Maximum Bit Rate* (kbps): conformant traffic follow a token bucket algorithm where token rate = Maximum bit rate and bucket size = Maximum SDU (Service Data Unit) size;

- *Guaranteed Bit Rate* (kbps): conformant traffic follow a token bucket algorithm where token rate = Guaranteed bit rate and bucket size = k^*Maximum SDU size with k = 1 (for release 1999);

- *Delivery Order* (y/n);

- *Maximum SDU Size* (octets);

- *SDU Format Information* (bits): list of exact sizes of SDUs;

- *SDU Error Ratio*: SDUs lost or detected as erroneous;

- *Residual Bit Error Ratio*: undetected bit error ratio (BER) in delivered SDUs;

- *Delivery of Erroneous SDUs* (y/n/-): 'yes' stands for 'erroneous SDUs detected and delivered with error indication'; 'no' stands for 'erroneous SDUs detected and discarded'; '-' stands for 'SDUs delivered without error detection';

- *Transfer Delay* (ms): maximum delay between Service Access Points for 95th percentile of delay distribution;

- *Traffic Handling Priority*: used to differentiate between bearer qualities;

- *Allocation/Retention Priority*: bearer relative importance for bearer allocation and retention;

- *Source statistics descriptor*: specifies characteristics of the source of submitted SDUs.

Table 2.3 lists the value ranges of the UMTS bearer service attributes. The value ranges reflect the capability of the UMTS network.

Table 2.3 Value ranges for UMTS bearer service attributes

Traffic class	Conversational class	Streaming class	Interactive class	Background class
Maximum bit rate (kbps)	$\leq 2048^{a,b}$	$\leq 2048^{a,b}$	$\leq 2048-$ overhead^{b,c}	$\leq 2048-$ overhead^{b,c}
Delivery order	Yes/No	Yes/No	Yes/No	Yes/No
Maximum SDU size (octets)	≤ 1500 or 1502^{d}	≤ 1500 or 1502^{d}	≤ 1500 or 1502^{d}	≤ 1500 or 1502^{d}
SDU format information	e	e		
Delivery of erroneous SDUs	Yes/No/-^{f}	Yes/No/-^{f}	Yes/No/-^{f}	Yes/No/-^{f}
Residual BER	$5 \times 10^{-2}, 10^{-2}$, $5 \times 10^{-3}, 10^{-3}$, $10^{-4}, 10^{-6}$	$5 \times 10^{-2}, 10^{-2}$, $5 \times 10^{-3}, 10^{-3}$, $10^{-4}, 10^{-5}, 10^{-6}$	$4 \times 10^{-3}, 10^{-5}$, 6×10^{-8g}	$4 \times 10^{-3}, 10^{-5}$, 6×10^{-8g}
SDU error ratio	$10^{-2}, 7 \times 10^{-3}$, $10^{-3}, 10^{-4}$, 10^{-5}	$10^{-1}, 10^{-2}$, $7 \times 10^{-3}, 10^{-3}$, $10^{-4}, 10^{-5}$	$10^{-3}, 10^{-4}$, 10^{-6}	$10^{-3}, 10^{-4}$, 10^{-6}
Transfer delay (ms)	100—maximum value	280^{h}—maximum value		
Guaranteed bit rate (kbps)	$\leq 2048^{a,b}$	$\leq 2048^{a,b}$		
Traffic handling priority			1, 2, 3	
Allocation/Retention priority	1, 2, 3	1, 2, 3	1, 2, 3	1, 2, 3
Source statistics descriptor	speech/unknown	speech/unknown		

^{a}Bit rate of 2048 kbps requires that UTRAN operates in transparent Radio Link Control (RLC) protocol mode, in this case the overhead from layer 2 protocols is negligible.

^{b}The granularity of the bit rate attributes is for further study. Although the UMTS network has capability to support a large number of different bit rate values, the number of possible values is limited so as not to unnecessarily increase the complexity of, for example, terminals, charging and inter-working functions.

^{c}Impact from layer 2 protocols on maximum bit rate in non-transparent RLC protocol mode shall be estimated.

^{d}In case of PDP type = Point-to-Point Protocol (PPP), maximum SDU size is 1502 octets. In other cases, maximum SDU size is 1500 octets.

^{e}Definition of possible values of exact SDU sizes for which UTRAN can support transparent RLC protocol mode is the task of radio access network Working Group 3.

^{f}If Delivery of erroneous SDUs is set to 'Yes' error indications can only be provided on the MT/TE side of the UMTS bearer. On the CN Gateway side error indications cannot be signalled outside of UMTS network in release 1999.

^{g}Values are derived from Cyclic Redundancy Check lengths of 8, 16 and 24 bits on layer 1.

^{h}If the UE requests a transfer delay value lower than the minimum value, this shall not cause the network (SGSN and GGSN) to reject the request from the UE. The network may negotiate the value for the transfer delay.

2.2.2.4 UMTS access technique, modulation and coding

Access Technique

The access scheme adopted is denoted as W-CDMA, with information spread over approximately 5 MHz bandwidth.

For what concerns the duplexing methods, two techniques are adopted:

- *Frequency Division Duplex*: This operates with paired bands on the uplink and downlink. Spreading factors are from 256 to 4 for FDD uplink, and from 512 to 4 with FDD downlink [ETS-03d].

- *Time Division Duplex*: This operates with unpaired bands. In this operating mode, the TDMA component is used, in addition to the W-CDMA technique. Spreading factors are from 16 to 1 for both TDD uplink and downlink [ETS-03c].

The radio frame is 10 ms long and is divided into 15 slots (2560 chip/slot at the chip rate 3.84 Mcps). The information rate of the channel varies with the symbol rate being derived from the 3.84 Mcps chip rate and the spreading factor. Thus, the respective modulation symbol rates vary from 960 ksymbols/s to 15 ksymbols/s (7.5 ksymbols/s) for FDD uplink (downlink), and for TDD the modulation symbol rates vary from 3.84 Msymbols/s to 240 ksymbols/s.

Channel Coding

Three options are supported:

- Convolutional coding;

- Turbo coding;

- No channel coding.

Bit interleaving is performed in order to pseudo-randomly distribute transmission errors.

Modulation

The adopted modulation scheme is QPSK.

In UTRA, different families of spreading codes are used to spread the signal, as detailed in [ETS-03f] for the FDD mode and [ETS-03c] for the TDD mode.

Channels separation from the same source:

- Walsh-Hadamard codes, also called Orthogonal Variable Spreading Factor codes. The generation of these codes is derived from a tree structure.

Separation from different cells:

- FDD mode: Gold codes with 10 ms period (38 400 chips at 3.84 Mcps) used, with the actual code itself of length $2^{18} - 1$ chips.

- TDD mode: Scrambling codes of length 16 are used.

Separation from different UEs:

- FDD mode: Gold codes with 10 ms period, or alternatively S(2) codes of 256 chip period.

- TDD mode: codes with period of 16 chips and mid-amble sequences of different length, depending on the environment.

2.2.2.5 User plane

The UMTS user plane consists of a layered protocol structure providing user information transfer, along with associated information transfer control procedures (e.g. flow control, error detection, error correction and error recovery). The user plane independence of the Network Subsystem platform from the underlying radio interface is preserved via the Gb-interface. Figure 2.13 depicts the user plane used in UMTS.

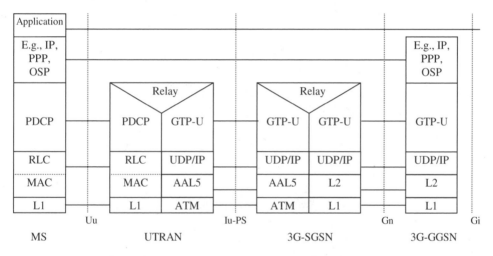

Figure 2.13 UMTS user plane protocol stack - PS domain.

In UMTS there is only the GTP-U layer above the UDP/IP layer. The Iu-PS interface connects the UTRAN and the PS domain of the CN. The user data and signalling information is transferred towards this interface. The protocol stack of the Iu-PS interface is UDP/IP over *ATM Adaptation Layer Type 5* (AAL5) over ATM. The transmitted data in the *Asynchronous Transfer Mode* (ATM) is divided into cells of size 48 octets. The AAL5/ATM part may use one or several permanent virtual circuits as common layer resources. The Uu-interface [ETS-03a] uses the Medium Access Control (MAC) functions such as ciphering and mapping between logical channels and transport channels. The RLC functions, which are located between the MAC layer and the higher layers perform error correction, ciphering, transfer of user data, segmentation and reassembly. The RLC layer provides the following modes: transparent mode, acknowledged mode and unacknowledged mode. In transparent mode the SDUs, which are received from the higher layers, are segmented into appropriate RLC PDUs without overhead. Moreover, the RLC PDUs are transferred to the MAC layer. The RLC layer also maps received PDUs from the MAC layer into SDUs for the higher layers. The *Packet Data Convergence Protocol* (PDCP) provides protocol transparency for higher-layer protocols and performs header compression, which improves the channel efficiency and header decompression of IP data streams at the transmitting and receiving entity. The PDCP SDUs are forwarded to the RLC layer and the RLC SDUs are transferred to the higher layers. Another function is the buffering of PDCP SDUs and the association of sequence numbers concerning to PDCP SDUs. The PDCP is located in the MS and the UTRAN. Moreover PDCP supports IPv4, PPP, Octet Stream Protocol and IPv6.

2.2.2.6 Control plane

Figure 2.14 depicts the control plane used in UMTS.

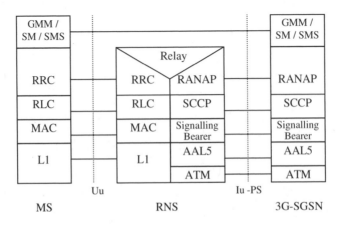

Figure 2.14 UMTS control plane protocol stack—PS domain.

The transport of signalling towards the Iu-interface is based on ATM/AAL5. The Signalling Bearer function enables the operator to choose between two different protocol options to transport SCCP (Signalling Connection Control Part) messages. The signalling bearer is located between the AAL5 layer and SCCP layers. The Radio Access Network Application Part (RANAP) is used for higher-layer signalling and is responsible for GTP connections on the Iu-PS-interface. The Radio Resource Control (RRC) layer carries out the establishment, the reconfiguration and the release of radio bearers, routing of higher layer PDUs and the establishment of an RRC connection between MS and UTRAN. GMM supports mobility management functions such as attach, detach and routing area updates and the *SM* supports PDP context procedures.

2.2.3 Wireless Local Area Network (W-LAN) Segment

2.2.3.1 Overview

The W-LAN segment provides short-range connectivity to prolong the satellite link in shadowed environments such as indoor or outdoor near-building areas, where the satellite

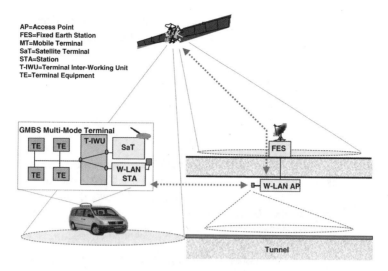

Figure 2.15 Satellite connectivity bridging via W-LAN.

coverage is poor or unavailable. Towards this end, suitably located M-ESW FESs are upgraded by integrating W-LAN equipment (see Figure 2.15).

The *W-LAN* system adopted in GMBS is based on the IEEE 802.11 standard [ISO-99]. The IEEE 802.11 belongs to the wide IEEE 802 family of standards [IEE-01] for local and metropolitan area networks. This family specifies physical and data link layers as defined by the International Organisation for Standardisation (ISO) Open Systems Interconnection (OSI) Basic Reference Model. The access standards belonging to the IEEE 802 family are all characterised by the same Logical Link Control (LLC) layer [ISO-98], while differing in the medium access technology and physical media. An 802.11 LAN is based on a cellular architecture where the system is subdivided into cells. Three basic topologies are envisaged:

- *The Independent Basic Service Set (IBSS)*: the IBSS configuration is an ad-hoc network containing a set of IEEE 802.11 stations that communicate directly with one another without using an access point (AP) or any connection to a wired network. This configuration is not used in GMBS.

- *The Basic Service Set (BSS)*: the BSS configuration consists of at least one AP connected to the wired network infrastructure and a set of wireless end stations. The BSS relies on an AP that acts as the logical server for a single W-LAN cell or channel. Communications between node A and node B actually flow from node A to the AP and then from the AP to node B. This configuration is used in the GMBS to cover single spot areas such as small underground parking, shadowed areas and so on.

- *The Extended Service Set (ESS)*: the ESS configuration consists of a series of overlapping BSSs (each containing an AP) connected together by means of a Distribution System (DS). Although the DS could be any type of network, it is almost invariably an Ethernet LAN. In some cases, the DS may be wireless. Mobile nodes can roam between APs while maintaining their communications path towards the network. This configuration is used in GMBS to cover large areas where satellite connectivity is unavailable, for example, a train station [LIA-03]. The whole interconnected W-LAN, including the different cells, their respective APs and the DS, is viewed as a single 802 network by the upper layers of the OSI model.

2.2.3.2   Network nodes

The network nodes envisaged in the W-LAN system are as follows:

- *Access Point*: this acts as the base station for the wireless network, aggregating access for multiple wireless stations onto the wired network. It usually consists of an IEEE 802.11 radio interface, a wired network interface (e.g. 802.3) towards the DS, and bridging software conforming to the 802.1d bridging standard.

- *Distribution System*: this represents the W-LAN backbone connecting several APs.

- *Station*: this is any device that contains IEEE 802.11 conformant MAC and physical layer interface to the wireless medium. A station is normally a PC equipped with a wireless network interface card: for instance 802.11 PC Card, PCI, or ISA NICs, or embedded solutions in non-PC clients (such as an 802.11-based telephone handset).

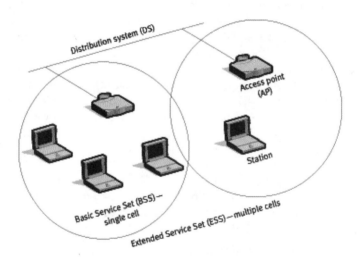

Figure 2.16   W-LAN architecture.

- *Portal*: this concept is defined in the IEE 802.11 standard. A portal is a device that interconnects between an 802.11 and another 802 LAN. This concept is an abstract description of part of the functionality of "a translation bridge". Even though the standard does not necessarily request it, typical installations will have the AP and the portal on a single physical entity.

The W-LAN segment does not have independent access to the F-ISP network. It is connected only to the M-ESW segment. The interfacing between the M-ESW and the W-LAN segments takes place in the upgraded M-ESW FESs. Depending on the dimension of the area to be covered, two typologies of wireless network configuration are envisaged:

- in the case of the BSS configuration for the W-LAN segment, a W-LAN AP is directly connected to the M-ESW FES;

- in the case of the ESS configuration for the W-LAN segment, a number of W-LAN APs are connected to the DS, typically an Ethernet LAN, and the DS is connected to the M-ESW FES.

2.2.3.3 User and control planes

In order to realise the interfacing between the M-ESW and W-LAN segment, suitable M-ESW FESs are upgraded by inserting an *Inter-working Layer* (IWL) operating:

- between the M-ESW layer 2-I (see Figure 2.4) and the IEEE 802.2 LLC layer (see Figure 2.17) in the user plane; IWL entities are seen by the IEEE 802.2 LLC entities as network layer entities;

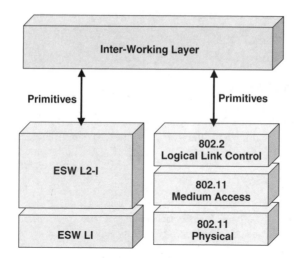

Figure 2.17 Upgraded M-ESW FES protocol stack: user plane.

- between the M-ESW layer 3 (connection control and location management entities, see Figure 2.5) and the IEEE 802.2 LLC (see Figure 2.18) in the control plane.

In the user plane, the IWL entities are in charge of executing all the operations necessary to carry out the exchange of data units between the M-ESW and W-LAN segments. This could include extraction/insertion of data units from/in the service primitives and setting of suitable service primitives' fields for addressing purposes.

In the control plane, the IWL entities are in charge of carrying out operations for the execution of location management, connection set-up and M-ESW/W-LAN handover procedures.

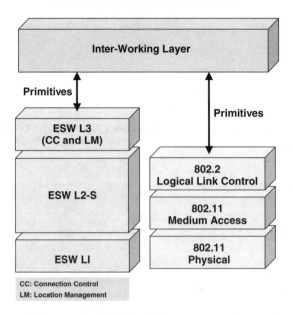

Figure 2.18   Upgraded M-ESW FES protocol stack: control plane.

ACKNOWLEDGEMENT

The authors wish to thank Dr Michele Luglio (Università di Roma "Tor Vergata") for his valuable contribution dealing with the non-geostationary satellite extension to the GMBS.

REFERENCES

[BET-99] C. Bettsetter, H-J Vögel, J. Eberspacher: GSM Phase 2+ General Packet Radio Service: Architecture, Protocols and Air Interface, *IEEE Communication Surveys & Tutorials*, **2**(3), Third Quarter 1999.

[ETS-01] ETSI TS 123 101 V4.0.0 (2001-04) Universal Mobile Telecommunications System (UMTS); General UMTS Architecture (3GPP TS 23.101 version 4.0.0 Release 4).

[ETS-02a] ETSI TR 125 931 V5.1.0 (2002-06) Universal Mobile Telecommunications System (UMTS); UTRAN Functions, Examples on Signalling Procedures (3GPP TR 25.931 version 5.1.0 Release 5).

[ETS-02b] ETSI TS 101 344 V7.9.0 (2002-09) Digital Cellular Telecommunication System (Phase 2+); General Packet Radio Service (GPRS) Service Description; Stage 2 (3GPP TS 03.60 version 7.9.0 Release 1998).

[ETS-03a] ETSI TS 125 223 V5.3.0 (2003-03) Universal Mobile Telecommunications System (UMTS); Spreading and Modulation (TDD) (3GPP TS 25.223 version 5.3.0 Release 5).

[ETS-03b] ETSI TS 123 002 V5.10.0 (2003-03) Digital Cellular Telecommunications System (Phase 2+); Universal Mobile Telecommunications System (UMTS); Network Architecture (3GPP TS 23.003 version 5.10.0 Release 5).

[ETS-03c] ETSI TS 125 222 V5.5.0 (2003-06) Universal Mobile Telecommunications System (UMTS); Multiplexing and Channel Coding (TDD) (3GPP TS 25.222 version 5.5.0 Release 5).

[ETS-03d] ETSI TS 125 212 V5.6.0 Universal Mobile Telecommunications System (UMTS); Multiplexing and Channel Coding (FDD) (3GPP TS 25.212 version 5.6.0 Release 5) (2003-09).

[ETS-03e] ETSI TR 125 922 V5.1.0 (2003-09) The Universal Mobile Telecommunications System (UMTS); Radio Resource Management Strategies (3GPP TR 25.922 version 5.1.0 Release 5).

[ETS-03f] ETSI TS 125 213 V5.4.0 (2003-09) Universal Mobile Telecommunications System (UMTS); Spreading and Modulation (FDD) (3GPP TS 25.213 version 5.4.0 Release 5).

[IEE-01] IEEE Std 802™-2001, Local and Metropolitan Area Networks: Overview and Architecture.

[ISO-98] ISO/IEC 8802-2:1998 (IEEE Std. 802.2™-1998) Information Technology—Telecommunications and Information Exchange between Systems—Local Metropolitan Networks - Specific Requirements—Part 2: Logical Link Control.

[ISO-99] ISO/IEC 8802-11:1999 (IEEE Std. 802.11™-1999) Information Technology—Telecommunications and Information Exchange between Systems—Local Metropolitan Networks - Specific Requirements—Part 11: Wireless LAN Medium Access Control (MAC) and Physical Layer (PHY) Specifications.

[LIA-03] X. Liang, F.L.C. Ong, P.M.L. Chan, R.E. Sheriff, P. Conforto: Mobile Internet Access for High-Speed Trains via Heterogeneous Networks, *Proceedings of 14th IEEE Personal, Indoor and Mobile Radio Communications (PIMRC 2003)*, **1**, 7–10 September 2003, Beijing, China; 177–181.

[MUR-00] R. Mura, G. Losquadro: A Satellite Network bringing Broadband Communications to the User, *Proceedings of 6th Ka Band Utilization Conference*, Cleveland, Ohio, USA, 31 May-2 June, 2000; 149–156.

[SAL-99] A.K. Salkintzis: A Survey of Mobile Data Networks, *IEEE Communication Surveys & Tutorials*, **2**(3), Third Quarter 1999.

ACRONYMS

AP	Access Point
BER	Bit Error Ratio
BG	Border Gateway
BGP	BG Protocol
BSC	Base Station Controller
BSS	Basic Service Set
CAC	Call Admission Control
CDMA	Code Division Multiple Access
CN	Core Network
CS	Code Scheme
DBAC	Dynamic Bandwidth Allocation Control
DS	Distribution System
E2E	End-to-End
EIR	Equipment Identity Register
ESS	Extended Service Set
FDD	Frequency Division Duplex
FES	Fixed Earth Station
GGSN	Gateway GSN
GMBS	Global Mobile Broadband System
GMM	GPRS Mobility Management
GMSC	Gateway MSC
GPRS	General Packet Radio Service

GSM	Global System for Mobile Communications
GSN	GPRS Support Node
GTP	GPRS Tunnelling Protocol
HLR	Home Location Register
HP	High Priority
HPLMN	Home PLMN
IBSS	Independent Basic Service Set
IMEI	International Mobile Equipment Identity
IMSI	International Mobile Subscriber Identity
ISP	Internet SP
IWL	Inter-Working Layer
IWMSC	Inter-Working MSC
LLC	Logical Link Control
LP	Low Priority
M-ESW	Mobile EuroSkyWay
MAC	Medium Access Control
MS	Mobile Station
MSC	Mobile Switching Centre
MT	Mobile Terminal
NAP	Network Access Point
NIU	Network Interface Unit
NOC	Network Operation Centre
NSP	Network Service Provider
OBP	On-Board Processing
PDCP	Packet Data Convergence Protocol
PDN	Packet Data Network
PDP	Packet Data Protocol
PDTCH	Packet Data Traffic Channel
PLMN	Public Land Mobile Network
PPP	Point-to-Point Protocol
PS	Packet Switched
QoS	Quality of Service
QPSK	Quadrature Phase Shift Keying
RANAP	RAN Application Part
RLC	Radio Link Control
RNC	Radio Network Controller
RNS	Radio Network Subsystem
RRC	Radio Resource Control
RS	Reed-Solomon
SaT	Satellite Terminal
SCCP	Signalling Connection Control Part
SDU	Service Data Unit
SGSN	Serving GSN
SM	Session Management
SN	Sub-Network
TDD	Time Division Duplex
TDMA	Time Division Multiple Access

TE	Terminal Equipment
TRM	Traffic Resource Management
UE	User Equipment
UMTS	Universal Mobile Telecommunications System
UT	User Terminal
UTRAN	UMTS Terrestrial Radio Access Network
VLR	Visitor Location Register
VPLMN	Visited PLMN
W-CDMA	Wideband CDMA

3

GMBS Multi-Mode Terminal

PAOLO CONFORTO, CLEMENTINA TOCCI

Alenia Spazio, Italy

3.1 GMBS MULTI-MODE TERMINAL

GMBS Multi-Mode Terminals (GMMTs) are applicable for both individual and collective use. Three different types are envisaged:

- car version, mounted on a personal vehicle;

- large vehicle version, mounted on collective means of transportation such as trains, buses and so on;

- briefcase version, specifically conceived for ease of transportation and fast service activation.

Table 3.1 summarises some of main features of the three terminal types in terms of:

1. where the terminal is mounted;

2. mobility, when in use and;

3. usage.

Moreover, an indication of the characteristics of the satellite terminal antenna is provided.

Figure 3.1 shows the internal structure of the *collective* GMMT. The diagram illustrates a number of Terminal Equipment (TE) connected to the Terminal Inter-Working Unit (T-IWU) via a Local Area Network (LAN). In the GMMT for *individual* use, a single TE directly connected to the T-IWU would replace the LAN. In the figure, the GPRS (General Packet Radio Service) and UMTS (Universal Mobile Telecommunications System) mobile terminals (MTs) are considered as alternative solutions for the implementation of the wireless terrestrial component; the W-LAN (Wireless Local Area Network) section is used to extend the connectivity of the satellite component in areas of poor satellite visibility. Table 3.2 lists the main blocks composing the GMMT and their functionality.

Space/Terrestrial Mobile Networks. Edited by R.E. Sheriff, Y.F. Hu, G. Losquadro, P. Conforto, C. Tocci
© 2004 John Wiley & Sons, Ltd   ISBN: 0-470-85031-0

Table 3.1 GMMT typologies

Type	Support	Mobility	Satellite Antenna	Usage
Car version	Car	Mobile	Quasi-planar antenna (to avoid compromising the aesthetic features of the car)	Collective
Large vehicle version	Collective means of transportation (e.g. trains, buses, ships etc.)	Mobile	Two types of antenna are envisaged. 1. Quasi-planar antenna: best solution 2. Protuberant antenna: acceptable	Collective
Briefcase version	Personal laptop	Portable	Two types of antenna are envisaged. 1. Fully flat antenna: preferred solution 2. Remotely deployable: acceptable	Individual/ Collective

The GMBS is able to support services with data rates belonging to one of the following three classes:

- *High Data Rate* (up to 512/2000 kbps): for composite multimedia services;

- *Medium Data Rate* (up to 150 kbps): for video-conferencing, and low quality video services;

- *Low Data Rate* (up to 64 kbps): for voice and data services.

Data rates of the different segment specific mobile terminals (SS-MTs) (i.e. M-ESW, GPRS, UMTS and W-LAN) operating in both the individual and collective configurations are summarised in Table 3.3 and classified on the basis of the mobility features and

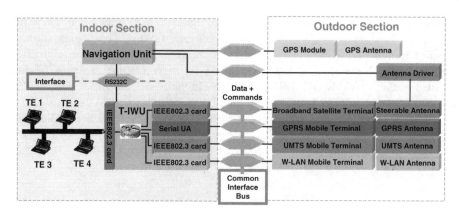

Figure 3.1 GMMT internal structure. Reproduced by permission from John Wiley and Sons Inc. © 2004.

Table 3.2 GMMT constituting block functionality

Component	Functionality
Broadband Satellite Terminal (SaT)	This is an M-ESW (Mobile EuroSkyWay) SaT operating at Ka-band (20/30 GHz). Three classes have been defined: (i) SaT-A (for both individual and collective use), (ii) SaT-B (for collective use) and (iii) SaT-C (for collective use). See Table 3.3 for details of the satellite terminals' data rate and operational environments.
Satellite Terminal Antenna	An active, phased-array antenna based on microstrip technology is adopted for all the satellite terminal typologies. A quasi-planar antenna, with a pointing mechanism that is electronic in elevation and mechanical in azimuth, is envisaged for mobile SaT-A, mobile SaT-B and mobile SaT-C terminals. The antennas for the three SaTs mainly differ in terms of Effective Isotropic Radiated Power (EIRP) and supported G/T (and consequently in size). A fully flat antenna, with a manual pointing mechanism, is envisaged for the portable SaT-A terminal.
	Alternatively, a protuberant antenna could be an acceptable solution (even though not optimal) for mobile SaTs mounted on large vehicles such as trains, buses, trucks etc. For the portable terminals, a remotely deployable antenna could represent a viable solution.
GPRS Mobile Terminal	The modem operates in the 900/1800 MHz bands and provides access to the GPRS radio resources. See Table 3.3 for details of data rate and operational environments.
UMTS Mobile Terminal	The modem operates in the 2 GHz band and provides access to the UMTS radio resources. See Table 3.3 for details of data rate and operational environments.
W-LAN Mobile Terminal	This encompasses an IEEE 802.11 modem operating at 2.4 GHz. See Table 3.3 for details of data rates and operational environments.
Terminal Inter-Working Unit	This hosts GMBS specific functionality for the management of inter-segment mobility and Quality of Service (QoS) over the access segments.
Terminal Equipment	The TE, also referred to as Mobile Host, is standard equipment (e.g. a laptop) which presents some upgraded specific functionality with respect to the commercial devices currently available. Two cases are envisaged:
	1. The TE is a laptop owned by the GMBS user. Such a user, once aboard the means of transportation (e.g. car, train, etc.), needs just a standard interface device to connect to the internal LAN and, through the LAN, to the T-IWU. In this case, the operating system of the TE is provided with a Mobile IP module.
	2. The TE is mounted onboard the means of transportation in the case where the GMBS user does not own one. The TE does not implement Mobile IP.
	In both cases, the TE implements RSVP (Resource Reservation Protocol) modules in order to request a certain level of QoS from the network. A GMBS Smart Card Reader is located at the TE; it reads the GMBS Smart Card, which stores the GMBS user profile (and therefore also the GMBS Subscriber Identity Module, GSIM) and is used for authentication and charging purposes.
Navigation Unit	This provides positioning data to:
	1. the pointing, acquisition and tracking (PAT) module to facilitate the calculation of the satellite antenna's azimuth and elevation;
	2. the T-IWU to support segment selection and handover procedures;
	3. the Internet applications (i.e. to the TE) for the deployment of location based services.

Table 3.3 Segment specific terminal data rate comparison

		Mobile			Portable	Service Area
		<10 km/h	<120 km/h	<250 km/h		
Satellite	Individual	Uplink: 160 kbps Downlink: 6 Mbps	Uplink: 160 kbps Downlink: 6 Mbps	N/A	Uplink: 160 kbps Downlink: 6 Mbps	Rural, Suburban/ Urban, Short range outdoor/indoor (via W-LAN), Polar zones (via LEO satellites)
	Collective	Uplink: 0.512/2 Mbps Downlink: 16 Mbps	Uplink: 0.512/2 Mbps Downlink: 16 Mbps	Uplink: 2 Mbps Downlink: 16 Mbps	Uplink: 0.512/2 Mbps Downlink: 16 Mbps	
GPRS	Individual	171.2 kbps	171.2 kbps	171.2 kbps	171.2 kbps	Suburban/ Urban, Short Range outdoor/indoor
	Collective[a]	171.2 kbps	171.2 kbps	171.2 kbps	171.2 kbps	
UMTS	Individual	2 Mbps	384/512 kbps	144/384 kbps	2 Mbps	Suburban/Urban, Short Range outdoor/indoor
	Collective[b]	2 Mbps	384/512 kbps[a]	144/384 kbps	2 Mbps	
W-LAN[c]	Individual	11/5.5/2/1 Mbps	11/5.5/2/1 Mbps	11/5.5/2/1 Mbps	11/5.5/2/1 Mbps[d]	Short Range outdoor/indoor
	Collective	11/5.5/2/1 Mbps	11/5.5/2/1 Mbps	11/5.5/2/1 Mbps	11/5.5/2/1 Mbps[d]	

[a]Higher bit rates may be obtained by connecting a number N of GPRS MTs to the T-IWU.
[b]Higher bit rates may be obtained by connecting a number N of UMTS MTs to the T-IWU.
[c]The W-LAN is used to bridge the satellite connectivity in shadowed areas.
[d]No speed limitation exists if the terminal remains in the same cell; stringent requirements for maximum speeds across cells' boundaries have to be met by the W-LAN product.

operational environments. Moreover, in the same table, the service area associated with each SS-MT is indicated.

As far as mobility support is concerned, it is envisaged that the GMBS supports:

- *terminal mobility*: services can be accessed through a GMBS terminal located anywhere within the GMBS service coverage area; the terminal can move during a service session at a speed of: (i) 250 km/h maximum in open/rural area environments; (ii) 120 km/h maximum in suburban/ urban environments; and (iii) 10 km/h maximum in indoor and low-range outdoor environments.

- *terminal portability*: services can be accessed through a GMBS terminal located anywhere within the GMBS service coverage area; the terminal would be static during a service session;

- *user mobility*: services (both mobile and portable) can be accessed by a GMBS user through any GMBS terminal while maintaining his/her own identity.

Tables 3.2 and 3.3 summarise the availability of the different access segments in the considered service environments, taking into account the degree of mobility and service rates on offer.

The GMMT's general architecture, depicted in Figure 3.1 for the case of collective use, can be particularised to the network scenario where the terminal has to operate. Considering the two major phases of the GMBS development, as discussed in Chapter 1, two main scenarios and consequently two GMMT typologies are envisaged:

- *Phase 1 GMMT:* the M-ESW SaT is complemented by only the GPRS MT; the W-LAN MT is also encompassed, which allows to bridge the satellite connectivity in some particular environments where the satellite coverage is not available;

- *Phase 2 GMMT:* the only difference with respect to the Phase 1 GMMT is that the UMTS MT complements or replaces the GPRS MT.

In both GMMT typologies the M-ESW SaT, the GPRS/UMTS MT and the W-LAN MT provide access to the wireless link resources guaranteeing the transparent transmission on the radio path of the IP packets generated by the IP-based TE.

Note: M-ESW SaT, GPRS/UMTS MT, W-LAN MT are also referred to as *SS-MTs*.

In the following sections, descriptions of the main architectural and functional characteristics of the GMMT components are provided.

3.2 TERMINAL INTER-WORKING UNIT

The T-IWU is involved in all the inter-segment mobility procedures and co-operates with the access segment specific mechanisms for the provision of QoS over the GMBS multi-segment access network.

The devised mobility management scheme foresees that the execution of specific inter-segment mobility procedures, such as segment selection, segment re-selection and inter-segment handover (ISHO), also known as vertical handover, is strictly related to the execution of specific Mobile IP (Internet Protocol) [IET-03] related tasks. The T-IWU is in charge of co-ordinating these activities in order to support the global mobility of the GMMT.

At the same time, the T-IWU is also responsible for the execution of specific tasks necessary for end-to-end QoS support. Toward this end, the T-IWU hosts suitable functional entities that provide, by interacting with their peers at the network side, users with the same QoS regardless of the access segment supporting the communication. These entities are grouped in a transparent layer placed between the underlying network layers and the IP layer and are in charge of executing QoS related functions (e.g. traffic shaping, policing, classification, scheduling and congestion control), which complement the mechanisms for the support of QoS envisaged in each access segment. These functional entities are selectively enabled/disabled, depending on the access segment.

The T-IWU functional block diagram is depicted in Figure 3.2. Each block represents a software module, grouping similar functionality executed inside the T-IWU.

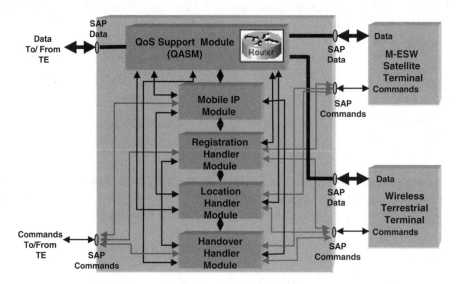

Figure 3.2   T-IWU functional block diagram. © IEEE 2002.

From the diagram, it can be seen that the T-IWU consists of five modules:
QoS Support module: This is in charge of

- performing inter-working between the access segment specific protocols and the RSVP [IET-97a, IET-97b] implemented to support an Integrated Service model on the edge portion of the GMBS;

- executing functions (e.g. traffic shaping, policing, classification, scheduling and congestion control) to complement the access segment specific mechanisms for the provision of QoS so as to render the GMBS multi-segment access network with a unique and uniform interface for QoS provision from an IP perspective.

Enhanced Mobile IP module: This implements Mobile IP [IET-03] functionality with enhanced capability to support a QoS aware, ISHO procedure. Such a procedure, requiring an integrated management of IP mobility and QoS, is executed when a change in the access segment supporting the active IP session is required. This can be due to either QoS related issues (e.g. when a new access segment becomes available that is more suited to the QoS needs of the user) or mobility related issues (when the GMMT detects that it is moving away from the coverage area of the currently serving access segment and is entering into the coverage area of a new one).

Registration Handler module: This hosts functionality that is mainly in charge of dealing with the execution of registration and authentication procedures. Such procedures occur at two different levels:

- at access segment level: each SS-MT executes a segment specific registration procedure towards the corresponding access network (see Chapter 7 for further details);
- at GMBS level: this is executed when the TE (with a GMBS Smart Card inserted) is connected to the T-IWU.

Location Handler module: This executes the procedures needed to locate user terminals and to establish IP connectivity. In particular, this module is in charge of

- executing inter-segment mobility procedures such as the segment selection/re-selection procedures, which results in the definition of the access segment to be used to support the IP data traffic;
- interacting with the SS-MTs in order to trigger segment specific procedures such as the satellite connection set-up or PDP (Packet Data Protocol) context activation, with the aim of establishing wireless connectivity.

Handover Handler module: This is in charge of dealing with the ISHO procedure. Three different phases are envisaged.

- Handover information gathering: the Handover Handler module collects both information concerning the status of the radio links from the SS-MTs and information regarding the QoS perceived by the user;
- Handover decision: the ISHO procedure is triggered based on the characteristics of the user profile and taking into account information concerning the congestion status of the access segment to which the call is to be handed over;
- Handover execution: the connection on the new access segment is set-up, the binding tables involved in the Mobile IP registration are updated and the resources on the old path are released.

Interactions among the different modules take place to trigger internal routines and pass the necessary parameters. In Figure 3.2, arrows linking the different modules illustrate these commands; bold arrows represent the flow of IP packets.

The interaction between the T-IWU and the SS-MTs is performed through two service access points (SAPs):

- *SAP for data transfer*: IP packets go through this access point;
- *SAP for commands*: specific commands (e.g. AT (Attention) commands in the case of GPRS/UMTS MTs) are exchanged through this access point in order to trigger segment specific procedures.

The T-IWU is able to translate all control message formats coming from TE(s)/T-IWU, which are used for terminal registration/deregistration and terminal session set-up/release purposes, into equivalent segment specific (i.e. M-ESW, GPRS, UMTS) message formats in order to correctly activate the segment specific procedures.

In complementary fashion, the T-IWU is able to translate the segment specific message formats into TE message formats in order to correctly process, at the user side, control messages coming from the network.

3.3   BROADBAND SATELLITE TERMINAL

The SaT envisaged in the M-ESW system is able to receive and transmit in the Ka-band (20/30 GHz). Three types of SaT are considered: SaT-A, SaT-B and SaT-C. Table 3.4 summarises their main features for the land mobile environment, while Figure 3.3 depicts the SaT connected to the T-IWU.

The SaT is composed of two main subsystems: (i) the *Outdoor Unit* (ODU) and (ii) the *Indoor Unit* (IDU).

The IDU comprises of the following modules:

- a Network Interface Unit (NIU): providing all the broadband satellite network access functionality;

- a Power Supply Unit: providing voltages and currents required by the IDU components and by the ODU system;

- a Man Machine Interface: including all the functionality needed to be used and monitored by the end-user.

The ODU comprises of the following modules:

- an IF interface;

- a RF front-end;

- a planar, phased-array antenna: from the different requirements defined for the SaT-A, SaT-B and SaT-C devices, and taking into account their operational environments, four different types of phased array SaT antenna[1] are envisaged:

  (i) a quasi-planar antenna for mobile SaT-A;

  (ii) a quasi-planar antenna for mobile SaT-B;

  (iii) a quasi-planar antenna for mobile SaT-C;

  (iv) a fully flat antenna for portable SaT-A.

---

[1] As highlighted in Table 3.4, a protuberant antenna could be acceptable for mobile satellite terminals mounted on-board collective means of transportation such as a train, bus, truck, etc. For the portable terminal a remotely deployable antenna could be an alternative, viable solution.

Table 3.4 Land mobile satellite terminal features

Terminal	Type-A		Type-B	Type-C
Usage	Individual/Collective		Collective	Collective
Antenna type	Fully flat for laptop	Quasi-planar for car	Quasi-planar for car; Protuberant acceptable for train, bus, truck etc.	Quasi-planar for car; Protuberant acceptable for train, bus, truck, etc.
Coverage volume		360° Azimuth / 20°–75° Elevation	360° Azimuth / 20°–75° Elevation	360° Azimuth / 20°–75° Elevation
User position with respect to antenna	Far	Far/Shielding	Far/Shielding	Far/Shielding
Pointing requirements	Manual	Electronic in elevation and mechanical in azimuth	Electronic in elevation and mechanical in azimuth for the quasi-planar antenna; Mechanical both in elevation and azimuth for the protuberant antenna	Electronic in elevation and mechanical in azimuth for the quasi-planar antenna; Mechanical both in elevation and azimuth for the protuberant antenna
Mobility	Portable	Mobile	Mobile	Mobile
Uplink max information rate	160 kbps	160 kbps	512 kbps	2 Mbps
Downlink max information rate	6 Mbps	6 Mbps	16 Mbps	16 Mbps
Receive Frequency	Ka-band	Ka-band	Ka-band	Ka-band
Transmit Frequency	Ka-band	Ka-band	Ka-band	Ka-band
G/T	5.1 dB/K	5.1 dB/K	11.7 dB/K	11.7 dB/K
EIRP	39.8 dBW	39.8 dBW	42.81 dBW	48.8 dBW

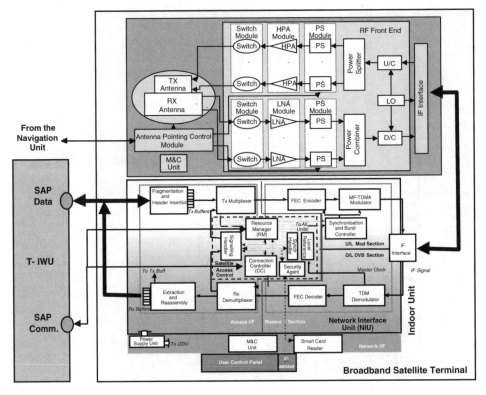

Figure 3.3   Broadband satellite terminal functional block diagram.

The same structure is adopted for the three kinds of mobile terminal antenna. The main differences lie in the EIRP and G/T requirements resulting in antennas of different dimensions and/or with a different number of amplifiers in the RF front-end.

- An internal Monitor and Control Unit (M&C Unit).

- An Antenna Pointing Control Module: steering (mechanical in azimuth and electronic in elevation) is envisaged for the mobile SaT-A, SaT-B and SaT-C antennas. An open loop PAT algorithm is used. Manual pointing is envisaged for the portable SaT-A antenna so, in this case, the Antenna Pointing Control Module is not present.

The ODU and IDU are interconnected via an *Inter Facility Link* (IFL) cable, which carries all the satellite signals (satellite signalling and both overlaying network traffic and signalling) to be exchanged between the IDU and the ODU subsystems and provides power to the ODU subsystem. The design is compatible with an IFL length of up to 100 m.
   On the basis of the SaT utilisation, the following installation scenarios can be considered:

- *Individual Installation*: in this case the SaT provides services to a *single GMBS user* (i.e. a single TE) connected to the IDU through the T-IWU;

- *Collective Installation*: in this case the IDU is connected, through the T-IWU, to a network (e.g. a LAN) to provide services to a set of GMBS users (i.e. a set of TEs). The M-ESW SaT types designed for these operating scenarios are preferably the SaT-B and SaT-C because of their higher supported information rate which allows the provision of satellite connectivity to a significant number of users; in the case of a small group of users (e.g. the passengers of a car), a SaT-A may also be adopted.

3.4 GPRS MOBILE TERMINAL

The main features of a GPRS mobile station (MS) are listed in Table 3.5.
The standard GPRS MS consists of two main blocks:

- the *Mobile Terminal* (MT) providing the access to the GPRS radio resources;

- the *Terminal Equipment* (TE) hosting IP and upper layer functionality.

Figure 3.4 depicts the GPRS MT connected to the T-IWU. The GPRS MT consists of three functional blocks corresponding to three main subsystems.

- *RF subsystem*: this filters and amplifies the signal received by the antenna. The transmitter generates, modulates and amplifies the signal to be transmitted. The receiver performs the inverse operation. A shift to an IF is required before baseband (BB) processing. A dual-band system is envisaged in the GPRS MT (GSM 900/DCS 1800).

- *BB Subsystem*: this can be divided into two sections: digital and analogue. The main element in the digital section is the BB Processor, which is in charge of the digital signal processing. A microprocessor is in charge of the man-machine interface and data services (communication ports). A set of memory chips plus all the interface logic required for a

Table 3.5 GPRS mobile station characteristics

Parameter	Value	Unit	Comment
Frequency Band (downlink)	935–960	MHz	GSM 900
Frequency Band (uplink)	890–915	MHz	GSM 900
Frequency Band Separation	20	MHz	
Duplex Spacing	45	MHz	
Number of Carriers (No. of RF Channels)	125	Carriers	Ch.0 → Ch.124
$N°$ of Guard Channels	1	Channel	Channel 0: 200 kHz
Carrier Spacing	0.2	MHz	Between two adjacent carriers.
Channel Bandwidth	0.2	MHz	
Number of Duplex Channels	124	Channels	Ch.1 → Ch.124
Channel Spacing	200	kHz	Between two adjacent carriers
Maximum Uplink Information Rate	171.2	kbps	Final version
Maximum Downlink Information Rate	171.2	kbps	Final version
Maximum Output Power (MS)	8 W (39 dBm)	W (dBm)	TRX. Power Class 2

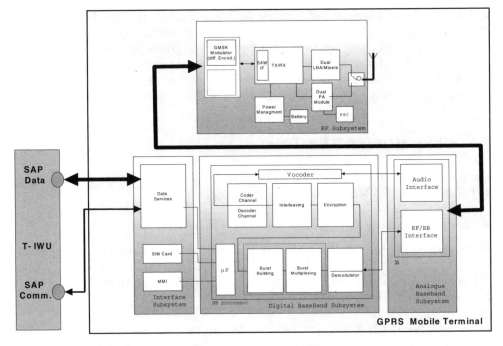

Figure 3.4   GPRS terminal functional block diagram.

GPRS module completes the functions performed by the BB processor. In the analogue section, the BB Analogue Interface performs functions relating to the audio system and other analogue parameters. A RF/BB Interface is required to perform adaptation between the RF Subsystem and BB Subsystem.

- *Interface subsystem*: the BB Processor commands this. The Subscriber Identity Module (SIM) Card functions and the Communication Ports are the main elements that make up this subsystem. Other interfaces, such as the display or keyboard, are envisaged.

GPRS may operate in an *individual configuration*, if only one TE is connected to the T-IWU, or a *collective configuration*, if several TEs are connected to the T-IWU through an internal LAN.

The interfacing between the T-IWU and GPRS MT takes place through the reference point R along with a second serial link at the same interface (see Figure 3.6 for equivalent UMTS interface).

A configuration may also be envisaged where several GPRS MTs are connected to the T-IWU (see Figure 3.5). In this case, several R interfaces are present in the same GMMT and all the specifications and considerations can be reapplied to each GPRS MT and each interface. Such a configuration would allow a maximum bit rate of $N \times 171.2$ kbps to be achieved, where N indicates the number of GPRS MTs installed in the GMMT.

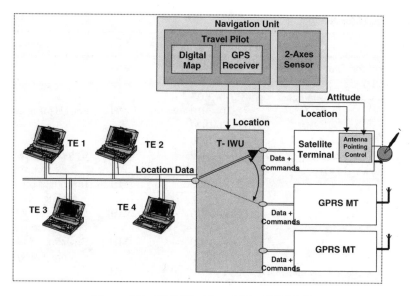

Figure 3.5   GMMT with multiple GPRS MTs.

3.5   UMTS MOBILE TERMINAL

The *User Equipment Domain* of the UMTS is divided into two domains: the *Mobile Equipment* (ME) *Domain* and the *User Service Identity Module* (USIM) *Domain*. The ME and the SIM make up the *Mobile Station* (MS).

The ME supports several combinations of the *Mobile Terminal* (MT), *Terminal Adapters* (TA) and *Terminal Equipment* (TE) functional groups. The MT part performs radio transmission, mobility management and session management.

The USIM contains information used for identification, authentication and ciphering of the subscriber. The SIM contains the same information as the USIM and so either the SIM or the USIM can be used in an UMTS MS.

The main features of the UMTS MS are listed in Table 3.6.

As depicted in Figure 3.6, the integration of the UMTS ME into the GMMT can be realised by introducing the T-IWU between the UMTS MT and the TE. Such a TE coincides with the TE described in Table 3.2. As with the satellite and GPRS operation, only one TE is connected to the T-IWU for the GMMT in the *individual configuration*, while, in the case of the *collective configuration*, several TEs can be connected to the T-IWU through an internal LAN.

As with GPRS, the interfacing between T-IWU and UMTS MT takes place through the reference point R.

Two fundamental functional blocks compose the UMTS MT:

- the *UMTS Terminal* strictly speaking, corresponding to the UMTS W-CDMA (Wideband Code Division Multiple Access) Terminal in which all UTRAN (UMTS Terrestrial Radio

Table 3.6 UMTS mobile station characteristics

Parameter	Value	Unit	Comment
Frequency band for Time Division Duplex (TDD) (uplink and downlink transmission)	1900–1920	MHz	
	2010–2025		
	1850–1910	MHz	For ITU Region 2
	1930–1990		
	1910–1930	MHz	For ITU Region 2
Frequency band for Frequency Division Duplex (FDD)	1920–1980	MHz	Uplink: Mobile transmits, Base receives
	2110–2170	MHz	Downlink: Base transmits, Mobile receives
	1850–1910	MHz	Uplink: Mobile transmits, Base receives
	1930–1990	MHz	Downlink: Base transmits, Mobile receives
Channel spacing (both TDD and FDD)	5	MHz	
Raster	200	kHz	
Error detection (both TDD and FDD)	CRC 24, CRC 16, CRC 12, CRC 8, CRC 0		See [ETS-03a] for TDD and [ETS-03b] for FDD
Coding scheme (both TDD and FDD)	Convolutional (1/2, 1/3) Turbo coding (1/3)		See [ETS-03a] for TDD and [ETSI-03b] for FDD
Maximum Uplink Information Rate	2000	kbps	
Maximum Downlink Information Rate	2000	kbps	
Mobile station maximum output power in TDD	21, 24	dBm	
Mobile station maximum output power in FDD	21, 24, 27, 33	dBm	

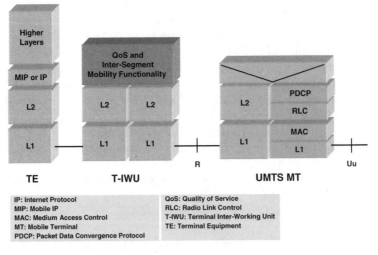

Figure 3.6 UMTS section general architecture description.

Access Network) functions are terminated, consisting of layers L1, MAC (Medium Access Control), RLC (Radio Link Control) and PDCP (Packet Data Convergence Protocol);

- the *Terminal Adapter* (also referred to as the UMTS-IP Terminal Adapter) implementing the terminal adaptation functions.

L1 and L2, illustrated in Figure 3.6 within the UMTS MT functional block, allow the UMTS MT and T-IWU to communicate with each other. A relay function is provided which allows exchange of information between the TA and the UMTS MT through standardised data formats.

The TA provides the T-IWU with the necessary command set to interact through the R reference point within the TA itself and ultimately with the UMTS network.

The UMTS MT consists of the following subsystems:

- RF subsystem;

- BB subsystem;

- Interface subsystem.

The above perform similar functions to their equivalent GPRS counterparts.

3.6 W-LAN MOBILE TERMINAL

The following blocks compose the W-LAN MS:

- the *W-LAN MT* providing access to the W-LAN radio channel;

- the *Terminal Equipment* hosting IP and upper layer functionality.

Table 3.7 IEEE 802.11x characteristics

Characteristic	IEEE 802.11	IEEE 802.11b	IEEE 802.11a
Spectrum Allocation	2.4 GHz	2.4 GHz	5 GHz
~Max physical rate	1.2 Mb/s	11 Mbit/s	54 Mb/s
~Max data rate, layer 3	1.2 Mb/s	5 Mb/s	32 Mb/s
Medium access Control/Media Sharing	CSMA/CA	CSMA/CA	CSMA/CA
Connectivity	Connectionless	Connectionless	Connectionless
Multicast	Yes	Yes	Yes
QoS support	Point Coordination Function (PCF)	PCF	PCF
Frequency selection	Frequency hopping Or Single carrier	Direct Sequence Spread Spectrum (DSSS)	DSSS
Encryption	40-bit RC4	40-bit RC4	40-bit RC4
Fixed network support	Ethernet	Ethernet	Ethernet
Management	802.11 MIB	802.11 MIB	802.11 MIB

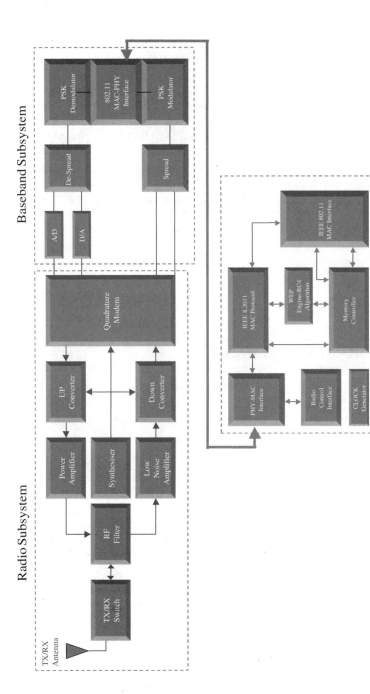

Figure 3.7 W-LAN terminal functional block diagram.

The W-LAN MT supports a transparent transmission of IP packets. The reference standard for the W-LAN MT architecture is IEEE 802.11, which defines the MAC sublayer and physical layer (whose main features are listed in Table 3.7) of the OSI Model. Three different transmission types characterise the physical layers: two spread-spectrum (Direct Sequence and Frequency Hopping) radio techniques and an infra-red specification. The MAC sublayer belongs to the family of CSMA/CA (Carrier Sense Multiple Access with Collision Avoidance) access protocols and defines two types of co-ordination function: peer-to-peer (Distributed Co-ordination Function) and centralised (Point Co-ordination Function).

Figure 3.7 depicts the W-LAN MT, which consists of three functional blocks corresponding to three main subsystems:

- the *Radio Subsystem*, filters and amplifies the received signal from the antenna. In the transmission phase, it generates, modulates and amplifies the signal to be relayed. The receiver performs the inverse operation. In both processes, the signal is translated in frequency by means of the UP/DOWN Converter;

- the *Baseband (BB) subsystem*, converts, modulates and spreads the received signal from the Radio and MAC subsystems;

- the *Medium Access Control (MAC) subsystem* tunes the radio channel access by way of CSMA/CA mechanism and implements the interfaces with the BB subsystem and IEEE 802.2 standard.

Figure 3.8 shows the W-LAN MT integration inside the GMMT.

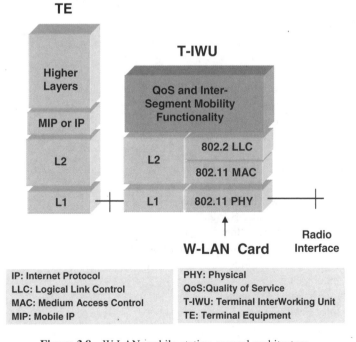

Figure 3.8 W-LAN mobile station general architecture.

3.7   NAVIGATION UNIT

The navigation unit supports:

- the antenna pointing of the SaT;
- Internet applications which require information of the user location and/or time;
- the execution of inter-segment selection and handover procedures.

The operation of the navigation unit is discussed further in Chapter 8.

REFERENCES

[ETS-03a] ETSI TS 125 222 V5.5.0 (2003–2006) Universal Mobile Telecommunications System (UMTS); Multiplexing and Channel Coding (TDD) (3GPP TS 25.222 version 5.5.0 Release 5).

[ETS-03b] ETSI TS 125 212 V5.6.0 Universal Mobile Telecommunications System (UMTS); Multiplexing and Channel Coding (FDD) (3GPP TS 25.212 version 5.6.0 Release 5) (2003–2009).

[IET-03] D.B. Johnson, C. Perkins, J. Arkko: Mobility Support in IPv6, *Internet Engineering Task Force*, Internet Draft draft-ietf-mobileip-ipv6-24.txt, Work in Progress, 30 June 2003.

[IET-97a] R. Braden (Ed.) *et al.*: Resource ReSerVation Protocol (RSVP)—Version 1—Functional Specification, *Internet Engineering Task Force*, RFC 2205, September 1997.

[IET-97b] J. Wroclawski: The use of RSVP with IETF Integrated Services, *Internet Engineering Task Force*, RFC 2210, September 1997.

ACRONYMS

AT	Attention
BB	Baseband
CSMA/CA	Carrier Sense Multiple Access with Code Avoidance
DCS	Digital Communication System
DSSS	Direct Sequence Spread Spectrum
EIRP	Effective Isotropic Radiated Power
FDD	Frequency Division Duplex
GMBS	Global Mobile Broadband System
GMMT	Global Multi-Mode Terminal
GPRS	General Packet Radio Service
GPS	Global Positioning System
GSIM	GMBS Subscriber Identity Module
GSM	Global System for Mobile Communications
HPA	High Power Amplifier
IDU	Indoor Unit
IF	Intermediate Frequency
IFL	Inter Facility Link
IP	Internet Protocol
ISHO	Inter-segment Handover

LAN	Local Area Network
LEO	Low Earth Orbit
LNA	Low Noise Amplifier
M & C	Monitor & Control
MAC	Medium Access Control
MS	Mobile Station
MT	Mobile Terminal
NIU	Network Interface Unit
ODU	Outdoor Unit
PAT	Pointing, Acquisition and Tracking
PCF	Point Coordination Function
PDCP	Packet Data Convergence Protocol
PDP	Packet Data Protocol
PS	Power Splitter
QoS	Quality of Service
RLC	Radio Link Control
RSVP	Resource Reservation Protocol
SAP	Service Access Point
SaT	Satellite Terminal
SIM	Subscriber Identity Module
SS-MT	Segment Specific MT
T-IWU	Terminal Inter-working Unit
TA	Terminal Adapter
TDD	Time Division Duplex
TE	Terminal Equipment
USIM	Universal SIM
USIM	User Service Identity Module domain
UTRAN	UMTS Terrestrial Radio Access Network
UMTS	Universal Mobile Telecommunications System
W-CDMA	Wideband Code Division Multiple Access
W-LAN	Wireless Local Area Network

4

Service Requirements

VINCENZO MARZIALE, PAOLO CONFORTO

Alenia Spazio, Italy

4.1 SERVICE SCENARIOS

4.1.1 Service Coverage

The Mobile EuroSkyWay (M-ESW) system is able to provide world-wide coverage. Due to the coverage offered by a geostationary satellite, connectivity is available everywhere bar the North and South Poles and in shadowed/indoor environments.

The development of the M-ESW system foresees a phased implementation. In the first phase, coverage is to be provided to the Extended European Region (see Figure 4.2), consisting of Europe, Middle East, Mediterranean Africa and some former Soviet Union countries. Such coverage is guaranteed by two co-located geostationary (GEO) satellites operating at Ka-band and characterised by on-board processing (OBP) functionality, which is responsible for managing traffic resources and switching.

The M-ESW service area is complemented in urban or indoor areas, where the satellite signal is weak or not available due to shadowing, by the service areas of either cellular mobile terrestrial networks, i.e. GPRS (General Packet Radio Service) and UMTS (Universal Mobile Telecommunications System), or Wireless Local Area Network (W-LAN) systems. This complementation represents one of the main features of the Global Mobile Broadband System (GMBS), which guarantees within the GMBS service area an average link availability of 99.5% per year.

The world's surface is partitioned into seven GEO coverage areas (see Figure 4.1):

- Extended Europe (see Figure 4.2);
- North America (see Figure 4.3);
- South America (see Figure 4.4);
- West Asia (see Figure 4.5);
- East Asia and Oceania (see Figure 4.6);
- Pacific Ocean (see Figure 4.7);
- Africa (see Figure 4.8).

Space/Terrestrial Mobile Networks. Edited by R.E. Sheriff, Y.F. Hu, G. Losquadro, P. Conforto, C. Tocci
© 2004 John Wiley & Sons, Ltd ISBN: 0-470-85031-0

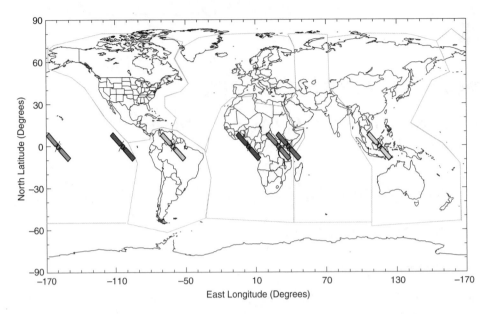

Figure 4.1   M-ESW world-wide service area coverage within the GMBS.

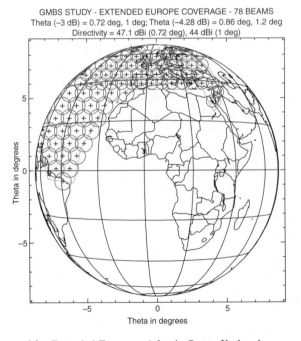

Figure 4.2   Extended Europe + Atlantic Ocean Ka-band coverage.

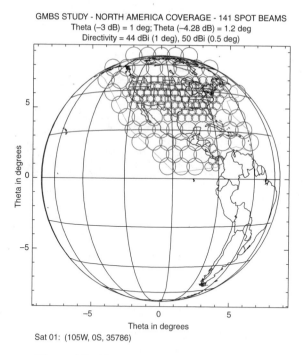

GMBS STUDY - NORTH AMERICA COVERAGE - 141 SPOT BEAMS
Theta (–3 dB) = 1 deg; Theta (–4.28 dB) = 1.2 deg
Directivity = 44 dBi (1 deg), 50 dBi (0.5 deg)

Sat 01: (105W, 0S, 35786)

Figure 4.3 North America Ka-band coverage.

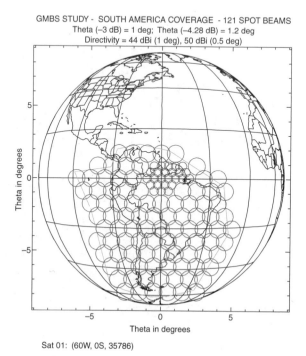

GMBS STUDY - SOUTH AMERICA COVERAGE - 121 SPOT BEAMS
Theta (–3 dB) = 1 deg; Theta (–4.28 dB) = 1.2 deg
Directivity = 44 dBi (1 deg), 50 dBi (0.5 deg)

Sat 01: (60W, 0S, 35786)

Figure 4.4 South America Ka-band coverage.

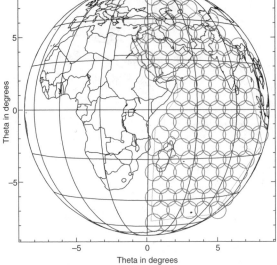

GMBS STUDY - WEST ASIA COVERAGE - 118 SPOT BEAMS
Theta (–3 dB) = 1 deg, 0.5 deg, Theta (–4.28 dB) = 1.2 deg, 0.6 deg
Directivity = 44 dBi (1 deg), 50 dBi (0.5 deg)

Sat 01: (39E, 0S, 35786)

Figure 4.5   West Asia Ka-band coverage.

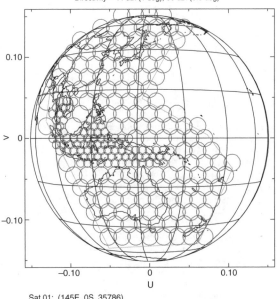

GMBS STUDY - EAST AISA AND OCEANIA COVERAGE - 240 SPOT BEAMS
Theta (–3 dB) = 1 deg, 0.5 deg; Theta (–4.28 dB) = 1.2deg, 0.6 deg
Directivity = 44 dBi (1 deg), 50 dBi (0.5 deg)

Sat 01: (145E, 0S, 35786)

Figure 4.6   East Asia and Oceania Ka-band coverage.

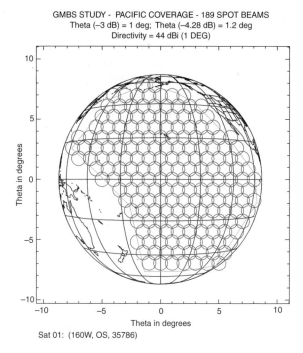

Figure 4.7 Pacific Ocean Ka-band coverage.

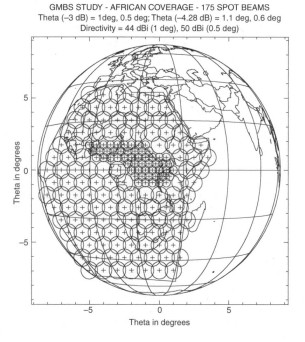

Figure 4.8 Africa Ka-band coverage.

4.1.2   Internet Service Type and Classification

An Internet service classification is not currently fixed in any standard. The classification envisaged in this Chapter is a combination of several sources and ongoing standards frameworks [ETS-00b, ETS-01, ETS-94, ETS-99, ITU-00a, ITU-93a, ITU-93b, ITU-94].

The Internet network considered in the GMBS is intended to support a wide set of ISP (Internet Service Provider) services that have been classified as

- *Client-Server Services*, i.e. the traditional Internet services, in this case with Quality of Service (QoS) guarantee;

- *Playback (Audio, Video) Services*, i.e. radio web-casting, TV web-casting and so on;

- *Multimedia Conversational Services*, i.e. video, audio and data conferencing, based on the T.120 [ITU-96], H.323 [ITU-00d] and SIP (Session Initiation Protocol) standards) [IET-02];

- *Supporting services* (e.g. Directory Services, DNS (Domain Name System)).

For each service class, a representative set of applications has been identified and, whenever considered necessary, one or more commercially available, reference applications have been indicated.

The application set for Client-Server Services is listed in Table 4.1.

The application set for Playback (Audio, Video) Services is listed in Table 4.2.

The application set for Multimedia Conversational Services is listed in Table 4.3.

The application set for Supporting Services is listed in Table 4.4.

Table 4.1   Application set for client-server services

Application	Example
Email	Microsoft Outlook Express, Microsoft Outlook, Eudora, Netscape Communicator
Ftp	
Telnet	
Web browsing	Microsoft Internet Explorer, Netscape Navigator

Table 4.2   Application set for playback (audio, video) services

Application	Example
Audio streaming: web-radio, web-casting	Real Video, Quick Time, Windows Media Player
Video streaming: web-TV, web-casting, video retrieval	Real Video, Quick Time, Windows Media Player

Table 4.3 Application set for multimedia conversational services

Application	Example
Data conferencing (T.120): program sharing, whiteboard, file transfer	NetMeeting™
Audio/video conferencing (H.323, SIP)	NetMeeting™

Table 4.4 Application set for supporting services

Application	Example
LDAP (Lightweight Directory Access Protocol) like services	ILS (Internet Locator Service): dynamic directory information, address book, user status (on-line/off-line), location, performance monitoring

The Internet Services in Tables 4.5–4.8 have been classified on the basis of six main attributes:

1. *Time dependence*: a real-time application requires strict delay and jitter (delay variation) constraints, while a non-real-time application, such as FTP (file transfer protocol), is mainly concerned with bandwidth constraints.

2. *Symmetry*: in asymmetric applications, requests are extremely less resource consuming than responses (e.g. the size of a request to download an MP3 file is a few bytes, while the file itself is some Mbytes). On the contrary, in symmetric applications, requests and responses are comparable in terms of resource consumption (e.g. each participant in a video conference would be both actor and audience).

3. *Connectivity*: a point-to-point connection involves two entities in a communication, while in point-to-multi-point connections, a sender communicates to many receivers.

4. *Rate*: the consistent availability of a fixed quantity of bandwidth is considered appropriate for Constant Bit Rate (CBR) services. For Variable Bit Rate (VBR) services, sources are expected to transmit at a rate which varies over time, i.e. the source can be described as "bursty". Finally, Available Bit Rate (ABR) services; many applications have the ability to reduce their information transfer rate if the network requires them to do so. Likewise, they may wish to increase their information transfer rate if there is extra bandwidth available within the network. There may not be deterministic parameters because the users are willing to operate with unreserved bandwidth.

5. *Nature*: a store-and-forward (asynchronous) communication allows a data streaming to be received at some interval after it has been sent. On the contrary, a *conversational* (synchronous) communication does not allow for extended delays between sending and receiving a stream of data.

6. *User*: the term user indicates the "final entity" accessing and utilising the service (e.g. a human, a machine, a service provider and so on).

Table 4.5 Internet service classification—guaranteed client-server

Application category	Main attributes	Example	Application	Application protocols	Comments
Guaranteed Client-Server	Time dependence	Non-Real-time (NRT)	E-mail	SMTP, POP	Document-mail Audio-mail, Voice messaging Video-mail
	Symmetry Connectivity Rate	Highly asymmetric Point-to-Point VBR, ABR	File transfer	FTP, TFTP	Data retrieval, Download (MP3 Audio files, MPEG4 Video files)
	Nature	Classical Client-Server Store and forward	Telnet Web browsing Enhanced browsing	TELNET HTTP Web S-HTTP	Remote Log-in, Server access Cached (Non-Interactive) Interactive (JAVA Applets...) Secure e-Commerce
	User	Either a machine or a human			

Table 4.6 Internet service classification—playback (Audio, Video)

Application category	Main attributes	Example	Application	Application protocols	Comments
Playback (Audio, Video)	Time dependence	Real-time (RT)	Audio streaming	RTP, RTCP	Radio web-casting
		No requirements on low transfer delay			Music web-distribution
		Limited delay variation (with respect to buffer)	Video streaming	RTP, RTSP, RTCP	TV web-casting
	Symmetry	Highly asymmetric	Audio broadcasting	MULTICAST	Video web-distribution
	Connectivity	Point-to-Point, Point-to-Multi-point	Video broadcasting	MULTICAST	Digital audio broadcast
	Rate	CBR, VBR			Digital video broadcast
	Nature	Store-and-forward			
	User	Live (human)			

Table 4.7 Internet service classification—multimedia conversational

Application category	Main attributes	Example	Application	Application protocols	Comments
Multimedia Conversational	Time Dependence	Real-time (RT)	Voice	RTP, RTCP	VoIP
		Low transfer delay requirements	Videoconference	RTP, RTCP	Live video, Videoconferencing
		Low delay variation requirements	Dataconference	T.120	Program sharing
	Symmetry	Symmetric			
	Connectivity	Point-to-Point, Point-to-Multi-point	Web games		Whiteboard
	Rate	CBR, VBR, ABR	Chat		Remote multi-players
	Nature	Conversational	Instant messaging unified multimedia messaging		
	User	Peers (or groups) of (human) users			

Table 4.8 Internet service classification—supporting services

Application category	Main attributes	Example	Application	Application protocols
Supporting services	Time Dependence	NRT	Domain name server	DNS
	Symmetry	Symmetric	Network time protocol	NTP
	Connectivity	Point-to-Point, Point-to-Multi-point	Dynamic host control Protocol	DHCP
	Rate	ABR, VBR	Lightweight directory access protocol	LDAP
	Nature	Signalling		SIP
		Call control		H.323
		Network management	Radius	RADIUS
	User	ISPs, Service providers, Content providers		

4.2 QoS SUPPORT

4.2.1 Supported Services

The GMBS is able to support all the services currently available to the fixed Internet community, such as

1. *Client-Server Services* (i.e. the traditional Internet services but with added functionality to provide QoS guarantees),

2. *Playback Services* (i.e, radio web-casting, TV web-casting, etc.),

3. *Multimedia Conversational Services* (i.e. video, audio, data conferencing based on the ITU-T T.120 and H.323, and Internet Engineering Task Force (IETF) SIP standards),

4. *Supporting Services* (e.g. Directory *Services*, DNS).

By using a *GMBS Multi-Mode Terminal* (GMMT), a GMBS user can activate these services by connecting to the Internet network. The components of the access network ensure that this can be achieved irrespective of the current location. The most suitable segment (satellite or terrestrial) to support access to the Internet is automatically or manually selected, this being dependent upon several factors such as radio coverage, operating environment, type of service and cost effectiveness.

Irrespective of the selected access segment, a GMBS user is allowed to declare and to negotiate with the GMBS the QoS required for the Internet service to be activated. Similarly, on the network side, the GMBS is able to provide hard guarantees for the Internet QoS negotiated with a GMBS user. Due to economic considerations, the GMBS user may also prefer not to have any guarantee for the QoS of the Internet services he/she is interested in.

The QoS requirements for the above services are in accord with the ITU (International Telecommunication Union), IETF and IMTC (International Multimedia Telecommunication

Consortium) standards. The GMBS provides network performance suitable for the support of these services.

The service performance requirements have been defined according to the ITU-T Recommendation I.350 [ITU-93a] network performance framework. The QoS evaluation approach, hereinafter adopted, employs a 3×4 matrix, obtained from the 3×3 matrix given in [ITU-93a] by adding a fourth column, to include the *Security* Performance Criterion, the others being Speed, Accuracy and Dependability. The rows of the matrix comprise of the basic communication functions, Access, User Information Transfer and Disengagement. It is worth noting that while it can be relatively easy to define the IP requirements for the information transfer phase, in general, it remains difficult to determine significant figures for the access and disengagement phases (see following section). For this reason, in the following, in some cases only the level of importance of the performance criteria of the access and disengagement phases is stated. A description of the rows and columns of the 3×4 matrix is provided in the following sections.

4.2.2   Description of the Basic Communication Functions

4.2.2.1   Access

The *access* function begins upon the issue of an access request signal or its implied equivalent at the interface between a user and the communication network. It ends when either:

1. a ready for data or equivalent signal is issued to the calling user; or

2. at least one bit of user information is input to the network (after connection establishment in connection-oriented services).

It includes all activities traditionally associated with physical circuit establishment (e.g. dialling, switching and ringing), as well as any activities performed at higher protocol layers.

4.2.2.2   User Information Transfer

The *user information transfer* begins on completion of the access function, and ends when the disengagement request terminating a communication session is issued. It includes all formatting, transmission, storage, error control and media conversion operations performed on the user information during this period, including necessary retransmission within the network.

4.2.2.3   Disengagement

There is a *disengagement* function associated with each participant in a communication session: each disengagement function begins on issue of a disengagement request signal. The disengagement function ends, for each user, when the network resources dedicated to that user's participation in the communication session have been released. Disengagement includes both physical circuit disconnection (when required) and higher-level protocol termination activities.

4.2.3 Description of Performance Criteria

4.2.3.1 Speed

Speed is the performance criterion that describes the time interval that is used to perform the function or the rate at which the function is performed (the function may or may not be performed with the desired accuracy).

4.2.3.2 Accuracy

Accuracy is the performance criterion that describes the degree of correctness with which the function is performed (the function may or may not be performed with the desired speed).

4.2.3.3 Dependability

Dependability is the performance criterion that describes the degree of certainty with which the function is performed regardless of speed or accuracy, but within a given observation interval.

4.2.3.4 Security

Security is the performance criterion that describes the degree of confidentiality with which the function is performed regardless of speed, accuracy or dependability.

4.2.4 Evaluation of QoS Requirements

4.2.4.1 The Quantitative-Qualitative Approach

To the best of the authors' knowledge, in the IP context, there are no suggestions to determine figures for the ITU-T I.350 [ITU-93a] attributes (i.e. speed, accuracy and dependability). This is mainly due to the fact that best effort (BE) does not allow any control over the actual delivered QoS.

For this reason, it is proposed to determine these values on the basis of the following criteria:

- *(C1) Current QoS performances*, achieved on BE IP networks through 56 kbps dial-up connections, *must* represent a *lower bound* to GMBS QoS requirements;

- *(C2) Application QoS requirements*, some applications, to function correctly, impose strict QoS requirements, e.g. to support an audio/video conference with a 6-inch picture size, a frame rate of 30 fps (frames per second) and audio toll quality, would require at least 384 kbps;

- *(C3) Wireless technology limits* impose an *upper bound* on QoS performance, e.g. a GPRS link can support at most about 150 kbps, while a M-ESW link can support in the return[1] direction up to 2 Mbps;

- *(C4) User expected performance* is *subjective*, which should be taken into account in the definition of GMBS QoS requirements.

To clarify the above concepts, consider a video streaming session. As noted above, the offer of video streaming with a 6-inch picture size, 30 fps and audio toll quality, implies that at least a 384 kbps link would be required (see C2). For some users, a video streaming with a 6-inch picture size, 15 fps and audio toll quality may be sufficient. In this case, a 128 kbps connection speed would be needed (see C4). In any case, a video streaming over GPRS cannot support a connection speed greater than 150 kbps, while the satellite can support up to 2 Mbps speed connection (see C3).

Consider another scenario: an ftp session to download a 3MB MP3 file. A 56 kbps wired connection allows the file to be downloaded in about 6 min. On the other hand, using a 2 Mbps M-ESW connection, the file can be downloaded in only 12 s. For an average user, it may be enough to download the MP3 file in 3 min, which means a download speed of 128 kbps.

Criteria (C1) and (C2) can be considered as *MUST requirements*, while criteria (C3) and (C4) can be considered as *SHOULD requirements*.

Table 4.9   Qualitative description of speed performance criteria

H	M	L
The function must be completed as fast as possible to guarantee good usability.	The function must be completed within a limited amount of time to guarantee satisfactory usability.	The function is not sensitive to speed performance.

Table 4.10   Qualitative description of accuracy performance criteria

H	M	L
High quality and correct data.	Partially damaged and/or medium quality data, the information must be easily intelligible.	Damaged and/or low quality data, the information must, however, be intelligible.

Whenever it is not possible to define a figure to specify the speed, accuracy and dependability performance criteria, the qualitative description specified in Tables 4.9–4.11 will be adopted.

For security performance criteria, there is no association with a numerical value to high (H), medium (M) and low (L) security level requirements.

---

[1]The return direction is the direction from the broadband satellite terminal towards the satellite Fixed Earth Station (FES). The 2 Mbps limit refers to the link from the satellite terminal to the satellite (i.e. uplink).

Table 4.11 Qualitative description of dependability performance criteria

H	M	L
99.9%	99%	95%

4.2.4.2 Common Requirements

Access speed is intrinsically limited by two factors: *DNS delay*, i.e. the time needed to obtain the IP address corresponding to the URL (Uniform Resource Locators), and; *Connection delay*, i.e the time needed to establish the TCP/IP connection. The access speed needs to satisfy the following requirements:

(C1) The sum of DNS delay and Connection Delay (denoted by AD in the following) must be lower than 1.5 s.

(C4) From a user point of view, if a communication is stable over time, namely the probability of premature *disconnection* is very low, the Access and Disengagement phase performances do not significantly affect the QoS experienced by the users. For example, a user may accept an access phase of a tele-conferencing service of several seconds if they are sure that this phase is performed once and with guaranteed good performance.

(C1) Login time, considering Modem negotiation, PPP (Point-to-Point Protocol [IET-94]) negotiation, PPP authentication, PPP Address assignment must be at most 35 s for 80% of the connections.

(C1) The probability of call failure, considering Busy signal, Ring no answer, Modem problem, Login failed must be lower than 2% for 80% of the connections.

(C3) The upper bounds imposed by the connection technologies are summarised in Table 4.12.

4.2.4.3 Network Performance Requirements for Internet Services

In this section the network performance requirements for some representative Internet services are reported.

Table 4.12 Access segment maximum transmission rate

System	Maximum transmission rate
GPRS	150 kbps
UMTS	144 kbps (high mobility), 512 kbps (low mobility), 2 Mbps (no mobility)
W-LAN	11 Mbps
M-ESW	160 kbps (SaT-A, uplink), 512 kbps (SaT-B, uplink), 2 Mbps (SaT-C, uplink)

Table 4.13   E-mail (SMTP (Simple Mail Transfer Protocol)) network performance requirements

Communication function		Performance criterion			
		Speed	Accuracy	Dependability	Security
E-mail (SMTP)	Access	(C1) AD[a] (C4) Access service delay[b] <5 s (90%)[c]	H	H	H[d]
	Information transfer	(C1) Empty mailbox check < 5 s (90%) (C1) Message delivery[e] < 5 min (90%)	H	H[f]	H
	Disengagement	L	H	H	L

[a]In this case, AD must be intended as the time to resolve the e-mail destination address.

[b]E-mail access delay is the interval between the click on the E-mail application icon and the retrieving of the first message in the mailbox.

[c]90% of the logins must take less than 5 s to be completed.

[d]If the SMTP server allows the forwarding function, the identity of the sender may not be guaranteed.

[e]At most 7% of the messages must be delivered after 1 h. The remaining 3% of the messages must be delivered in at most 12 h.

[f]POP3 receive failures less than 1–1.5%. Percentage of SMTP send failure less than 1%.

E-mail (SMTP) Network Performance Requirements

Network performance requirements for E-mail (SMTP) service are as listed in Table 4.13.

File Transfer Network Performance Requirements

Network performance requirements for File Transfer (FTP) service are as listed in Table 4.14.

Telnet Network Performance Requirements

Network performance requirements for Telnet service are as listed in Table 4.15.

Web Browsing (HTTP (Hypertext Transfer Protocol)) Network Performance Requirements

Network performance requirements for Web Browsing (HTTP) services are defined in terms of the following parameters:

- *DNS delay (Dd)*: the time needed to obtain an IP address corresponding to an URL;

- *Connection delay (Cd)*: the time needed to establish the TCP/IP connection;

- *Server delay (Sd)*: the time elapsed between the sending of the GET URL command and the reception of the first bytes of response;

- *Transmission delay (Td)*: this corresponds to the delay needed for the transmission of the remainder of the page, from connect to end;

- *Page delay (Pd)*: the sum of the transmission delay and the DNS delay;

Table 4.14 File transfer network performance requirements

	Communication function	Performance criterion			
		Speed	Accuracy	Dependability	Security
File transfer (FTP over TCP)	Access	(C1) AD^a	H	H	H
		(C4) Access delay^b <5 s (90%)			
	Information transfer	(C1) Average rate: 44 kbps	H	H	H
		(C4) Average rate: ≥128 kbps^c (80%)			
	Disengagement	L	H	H	L

^{a}In this case, AD must be intended as the time necessary to resolve the FTP/Telnet server address and to establish the connection to the FTP server. Since this procedure is performed once for each session, possible variations in AD time will affect users just once.

^{b}FTP access delay is the time elapsed between the sending of the FTP/Telnet access request to the server and the insertion of username and password.

^{c}This transfer rate allows downloading of 1 MB file in 1 min.

- *Transmission speed (Ts)*: the data rate of the connection;

- *Share of delay due to servers (Sds)*: the ratio of (server delay—round trip time) over (transmission delay).

Requirements for each of the above parameters are given in terms of 95th percentile[2], median (the 50th percentile) and average. They are listed in Table 4.16. All the figures in the table must be viewed as (C1) requirements.

Table 4.15 Telnet network performance requirements

	Communication function	Performance criterion			
		Speed	Accuracy	Dependability	Security
Telnet (over TCP)	Access	(C1) AD^a (C4) Access delay^b <5 s (90%)	H	H	H
	Information transfer	(C4) Response time <400 ms (90%)	M	H	H
	Disengagement	L	H	H	H

^{a}FTP access delay is the time elapsed between the sending of the FTP/Telnet access request to the server and the insertion of username and password.

^{b}This transfer rate allows downloading of a 1 MB file in 1 min.

[2]95th percentile of DNS delay equal to X, means that 95% of the requests have a DNS delay less than or equal to X.

Table 4.16 Web browsing (HTTP) network performance requirements

Communication function		Performance criterion			
		Speed[a]	Accuracy	Dependability	Security
Web browsing (HTTP over TCP)	Access	Dd 2191 ms/578 ms/ 5926 ms	H	H	
		Cd 844 ms/255 ms/ 6217 ms			
		Sd 1209 ms/338 ms/ 4761 ms			M[b]
	Information transfer	Td 2666 ms/988 ms/ 8219 ms			
		Ts 30.5 kbps/ 17 kbps	M	M	M[c]
	Disengagement	L	M	M	L[c]

[a]Average/median/95 percentile.
[b]A high level requirement should be considered for HTTP-S protocol based applications.
[c]The data rate is based on 12-inch monitor with a resolution of 640 × 480 pixels; picture size will change with increase or decrease in size of monitor.

Data Conferencing Network Performance Requirements

Network performance requirements for data conferencing service are as listed in Table 4.17.

Note: Data conferencing is based on the use of the T.120 protocol. ITU Recommendation T.120 [ITU-96] is made up of a suite of communication and application protocols. With T.120-based programs, multiple users can participate in conferencing sessions over different types of networks and connections.

Audio/Video Conferencing Network Performance Requirements

Network performance requirements for audio/video conferencing services are as listed in Table 4.18.

Note: Audio/Video conferencing is based on the use of the H.323 protocol. ITU Recommendation H.323 [ITU-00d] provides specification for computers, equipment, and services for multimedia communication over networks that do not provide a guaranteed QoS.

Table 4.17 Data conferencing (T.120) network performance requirements

Communication function		Performance criterion			
		Speed	Accuracy	Dependability	Security
Data conferencing (T.120)	Access	L	H	M	M
	Information transfer	Average rate: 128 kbps	M	H	M
	Disengagement	L	H	H	L

Table 4.18 Audio/video conferencing (H.323) network performance requirements

	Communication function	Performance criterion			
		Speed	Accuracy	Dependability	Security
Audio/Video	Access	L	H	H	M
conferencing	Information	Average Rate:	M	H	M
(H.323)	transfer	384 kbps			
	Disengagement	L	M	M	L

This standard is based on the IETF Real-Time Transport Protocol (RTP) [IET-96] and Real-Time Control Protocol (RTCP), with additional protocols for call signalling and data and audiovisual communications.

4.2.4.4   Additional Video and Audio Requirements

(C2) Digital Video Requirements

As far as digital video requirements are concerned, the H.323 family of standards [ITU-00d] defines three basic picture sizes in pixel density:

- Full Common Interface Format of 352-by-288 pixels;
- Quarter Common Interface Format (QCIF) of 176-by-144 pixels;
- Sub QCIF of 128-by-96 pixels (thumbnail size).

Several basic alternatives for video traffic demand are defined ranging from about 1 kbps up to about 1 Mbps.

Video virtually never travels without audio. The alternative audio standards must therefore be added to the basic video rates. Here, the audio quality more closely matches radio rather than telephone. The industry has settled upon two fundamental rates: 16 kbps, similar to AM radio (or telephone); and 64 kbps, which is similar to FM radio. Adding these two different rates to the basic video rates doubles the traffic loads to be considered. Table 4.19 reports some service data rates for audio/video services characterised by different picture sizes, frame rates and audio qualities.

(C2) Audio and Video Streaming Requirements

In the case of video and audio streams, the same requirements for H.323 [ITU-00d] services in terms of bandwidth and delay are applicable.

VoIP QoS Requirements

Physical Components Needed for VoIP Service Provision

The following components may be present in a VoIP system and each may contribute to the overall end-to-end (E2E) QoS performance of the system:

Table 4.19 Service data rate for audio/video services (H.323)

Digital audio and video data rate (kbps)[a]	Picture size (inches diameter)	Frame rate (fps)	Audio quality
28.8	2	1	Fair
128	2	20–30	Toll quality or better
128	6	10–15	Toll quality or better
384	6	30	Toll quality or better

[a]The data rate is based on 12-inch moniter with a resolution of 640×480 pixels; picture size will change with increase or decrease in size of monitor.

- IP terminal, intended for connection to an IP network;

- IP access network, a variety of access networks used to interconnect an IP terminal with IP backbone networks;

- IP backbone, a variety of equipment used to provide IP backbone networks;

- IWF (Inter-Working Function, gateway/gatekeeper), entities in charge of controlling, signalling and protocol conversion;

- Circuit-Switched Network;

- A variety of voice telephony terminals interconnected to various branches of the circuit-switched network.

Main QoS Parameters Influenced by VoIP System

E2E QoS in a VoIP system is characterised in the ETSI Project TIPHON (Telecommunications and Internet Protocol Harmonization over Networks) QoS under two broad headings:

- call set-up quality;

- call quality.

Call set-up quality is mainly characterised by the call set-up time, which is perceived by the user as the responsiveness of the service. Call set-up time is the time elapsed from the end of the user interface command by the caller (keypad dialling, e-mail alias typing, etc.) to the receipt by the caller of meaningful progress information. Call quality is characterised by the overall transmission quality rating R, which describes the full acoustic-to-acoustic (mouth-to-ear) quality, experienced by a user, for a typical situation using a 'standard' telephony handset. The overall transmission quality rating is calculated using the E-Model (see [ITU-00b]). For calculation purposes, the use of traditional telephone handsets (see [ITU-00c]) at both sides of the connection is assumed.

Within the overall transmission quality, two major factors contribute to the overall QoS of the VoIP system experienced by the user:

- *end-to-end delay*: this mainly impacts on the interactivity of a conversation. The measurement is performed from the mouth of the speaker to the ear of the listener;

- *end-to-end speech quality*: this is the one-way speech quality as perceived in a non-interactive situation.

The measurement methodologies for these parameters are specified in [ETS-02a], while the requirements for these parameters with respect to the various VoIP QoS classes are defined in [ETS-02b].

(C2) VoIP Specific QoS Factors

Examples of VoIP specific QoS relevant factors are:

- number of hops;
- possible variation of the geographical length of one connection during the talking state;
- occurrence of congestion;
- use of prioritisation or bandwidth reservation schemes.

VoIP Classes

Four classes of E2E QoS are defined for VoIP systems. The TIPHON QoS definitions include both the network and the VoIP terminal characteristics:

- *BEST*: This is a type of IP telephony service that has the potential to provide a user experience better than the PSTN (Public Switched Telephone Network).
- *HIGH*: This is a type of IP telephony service that has the potential to provide a user experience similar to PSTN.
- *MEDIUM*: This is a type of IP telephony service that has the potential to provide a user experience similar to common wireless mobile telephony services, for instance GSM networks using FR (Full Rate) codecs.
- *BEST EFFORT*: This type of service will provide a usable communications service but may not provide guarantees of performance.

Note: Connections that include a geostationary satellite will incur long propagation delay and consequently will fall into the BE QoS class. However, the user may experience QoS characteristics, other than E2E delay, corresponding to higher classes.

For the purposes of voice quality, each of the above classes is defined by three performance metrics:

- *Overall Transmission Quality Rating (R)*;
- *Listener Speech Quality* (one-way non-interactive E2E speech quality);
- *End-to-end* (mean one-way) *Delay*.

For a VoIP system to be considered as achieving a specified QoS class, it must meet all the three specified performance metrics for that particular class, for 95% of all connections.

Note 1: The *R*-value incorporates all degradations, including the effects of packet loss.

Table 4.20   VoIP QoS classes (ETSI TIPHON)

End-to-End QoS classes		4 (Best)	3 (High)	2 (Medium)	1 (Best Effort)
Intuitive QoS rating		> PSTN	~ PSTN	~ GSM	< GSM
Performance: metrics	Overall transmission quality rating (R)	Under study	< 85	< 70	< 50
	Listener speech quality	Better than G.711 [ITU-88]	Equivalent or better than G.726 [ITU-90] at 32 kbps	Equivalent or better than GSM-FR	Not defined
	End-to-End delay	< 100 ms	< 100 ms	< 150 ms	< 400 ms

Note 2: The *R*-value characterisation of systems employing wideband codecs has yet to be investigated.

Note 3: The rating for the BE class is a target value and can be treated as a guaranteed service if the target value is achieved.

The relation between overall transmission quality rating (*R*) and user perception of quality is defined in [ITU-99c]. Table 4.21 is taken from this Recommendation.

Call Set-up Time

As guidance, [ITU-99b] recommends mean values for call set-up delay, as shown in Table 4.22.

Presently, no upper bound figures are specified.

Table 4.21   Category of speech transmission Quality as defined by ITU-T G.109[ITU-99c]

R Value Range	90 < R < 100	80 < R < 90	70 < R < 80	60 < R < 70	50 < R < 60
User's satisfaction	Very satisfied	Satisfied	Some users dissatisfied	Many users dissatisfied	Nearly all users dissatisfied

Table 4.22   Call set-up mean time

Local Call	< 3 s
Toll (National Long-Distance) call	< 5 s
International call	< 8 s

Overall End-to-End Delay Calculation

The overall E2E delay can be determined as the sum of:

- Speech coding and packetisation delay;
- Network routing delay;
- Propagation delay;
- Delay variation buffer size.

Definition of Network Classes

According to the Packet Loss percentage and the Delay Variation, the following VoIP network classes can be defined, as shown in Table 4.23.

Table 4.23   Definition of network classes

Network class	Packet loss	Delay variation
I	< 0.5%	< 10 ms
II	< 1%	< 20 ms
III	< 2%	< 40 ms

Definition of Terminal Modes

Three terminal "Modes" A, B and C are defined on the basis of delay to allow for different speech encoding and packetisation schemes, as shown Table 4.24. In this context, terminal delay includes encode/decode and packetisation operations, but not jitter compensation.

Taking the values of delay for the specific Network Class and Terminal Mode in Tables 4.23 and 4.24 and noting that:

$$(Network\ Routing\ Delay + Propagation\ Delay) = Overall\ Delay\ for\ the\ VoIP\ QoS\ Class \\ - (Speech\ Coding\ and\ Packetization \\ Delay + Jitter\ Compensation\ Delay)$$

where Jitter Compensation Delay shall at least be equal to the network delay variation, Table 4.25 shows the propagation and routing delay available to meet a given VoIP QoS Class.

Table 4.24   Definition of terminal modes

Terminal mode	delay	Application
A	< 50 ms	Allows for small speech frame duration and small number of frames per IP Packet
B	< 75 ms	Allows for large speech frame duration and/or small number of frames per IP Packet
C	< 100 ms	Allows for larger speech frame duration and multiple frames per IP Packet

Table 4.25   Available propagation and routing delay as a function of VoIP QoS class, terminal mode and network class

Network Class	QoS Class	Terminal mode		
		A	B	C
I	High	40	15	X
	Medium	90	65	40
	Best effort	340	315	290
II	High	30	5	X
	Medium	80	55	30
	Best effort	330	305	280
III	High	10	X	X
	Medium	60	35	10
	Best effort	310	285	260

4.3   SECURITY

4.3.1   Overall GMBS Security Requirements

As highlighted in the previous section, security is one of the performance criteria to be considered in the definition of the Internet service QoS requirements. In order to reach the level of security needed by the Internet services, a number of functional requirements have to be satisfied by the different GMBS components.

The GMBS aims to prevent intruders from reading or modifying the data being transmitted or stored, as well as from gaining access to the resources or services of the system.

The GMBS security architecture is based on a two-layer approach:

- *lower layer*: this hosts the set of security functionality implemented by the systems composing the GMBS multi-segment access network;
- *upper layer*: this hosts the set of IP related security functionality implemented in specific nodes of the Federated ISP (F-ISP) network.

In the following, the functional security requirements pertaining to the two layers of the security architecture will be given.

4.3.2   Lower Layer Functional Security Requirements

4.3.2.1   M-ESW

The main functional security requirements to be met by the M-ESW system are as follows:

- *Verification of identities*: M-ESW supports capabilities to establish and verify the claimed identity of any actor in the network.

- *Controlled access and authorisation*: M-ESW supports capabilities to ensure that actors are prevented from gaining access to information or resources they are not authorised to access.

- *Protection of confidentiality*: M-ESW supports the capabilities to keep stored and communicated data confidential. Protection of confidentiality applies to both (i) user related M-ESW network information and, (ii) information used by other security services, (e.g. cryptographic keys).

- *Protection of data integrity*: M-ESW guarantees integrity of stored and communicated data. Protection of data integrity applies to both: (i) M-ESW user related network information and, (ii) information used by other services.

- *Strong accountability*: M-ESW guarantees that an entity cannot deny the responsibility for any of its performed actions as well as their effects. Strong accountability requires that any entity in a network must hold full responsibility for any of its actions. In other words, the entity may not repudiate its actions in the network.

- *Activity logging*: the network supports the capability to retrieve information about security activities stored in the network elements with the possibility of tracing this information to individuals or entities.

- *Alarm reporting*: M-ESW supports the capability to generate alarm notifications about certain adjustable and selective security related events.

- *Audit*: M-ESW supports the capability to analyse and exploit logged data on security relevant events in order to check them upon violations of system and network security.

- *Security recovery, management of security*: M-ESW supports recovery from successful and attempted breaches of security. Whenever an attempt to breach security occurs, it is possible to handle this attempt in a controlled manner, meaning that the attempt does not result in a severe degradation of M-ESW network availability.

In order to meet the requirements defined above, M-ESW implements the following security services:

- *Authentication service*: three different typologies may be identified:

  user authentication: delivering corroboration of the user;

  peer entity authentication: establishing proof of the identity of the peer entity at one particular moment in time during a communication relationship;

  data origin: establishing proof of identity of the peer entity responsible for a specific data unit.

- *Access control service*: this service provides a means to ensure that the subject accesses resources only in an authorised manner. This protection service may be applied to various types of access to a resource (e.g. the use of a communications resource; the reading, the writing, or the deletion of an information resource; the execution of a processing resource) or to all accesses to a resource. The limitations of access to these resources are laid out in access control information, which specifies: (i) the means to determine which entities are authorised to have access to an object, and (ii) what kind of access is allowed. Granting access to objects requires verification of the identity of the entity trying to gain access.

- *Confidentiality service*: this service provides protection against unauthorised disclosure of exchanged data. Two typologies are envisaged:

  data confidentiality;

  M-ESW signalling message confidentiality.

- *Integrity service*: this service provides the means to ensure the correctness of exchanged data, protection against modification, deletion, creation (insertion) and replay of exchanged data.

- *Non-repudiation service*: this service provides the means to prove that exchange of data actually took place.

- *Security logging service*: this service logs information about security relevant events that have occurred or security relevant operations that have been performed or attempted.

- *Security alarm*: the security alarm notification provides information regarding operational conditions and QoS, pertaining to security.

- *Security audit trail service*: an audit is to be seen as an independent review and examination of system information and activities in order to test for adequacy of system controls, to ensure compliance with the established security policy and operational procedures, to detect breaches in security and to recommend in control, policy and procedures.

- *Recovery service*: this service provides the means to recover from successful and attempted breaches on security.

4.3.2.2 GPRS

The security features for the GPRS system are mainly focused on the protection of the radio path from possible eavesdroppers and intruders. The aim of the included security features is to ensure the user data and identity confidentiality and to avoid the unauthorised utilisation of the service from possible intruders.

The main functional security requirements to be met by the GPRS system are as follows:

- *subscriber identity confidentiality*;
- *subscriber identity authentication;*
- *user data confidentiality on physical connections;*
- *connectionless user data confidentiality;*
- *signalling information element confidentiality.*

In order to meet the above, GPRS implements the following security services:

- Subscriber identity is represented by a unique International Mobile Subscriber Identity (IMSI). Subscriber identity confidentiality (i.e. the protection of the identity privacy for the subscribers who are using GPRS) is ensured by not making the IMSI available or disclosed to unauthorised individuals, entities or processes. The confidentiality of the

IMSI when it is transferred in signalling messages should be ensured, together with specific measures to preclude the possibility to derive it indirectly from listening to specific information, such as addresses, at the radio path.

- An authentication mechanism to corroborate the subscriber identity, IMSI or TMSI (Temporary Mobile Subscriber Identity), at the land-based part of the system is included. The authentication security feature is able to protect the network against unauthorised use, enabling also the protection of the GPRS subscribers by denying the possibility for intruders to impersonate authorised users. The authentication procedure is invoked:

    whenever an access to a service (including some or all of: set-up of mobile originating or terminating calls, activation or deactivation of a supplementary service) is requested;

    whenever a change of a subscriber-related information element in the Home Location Register occurs.

- Functionality is implemented for the protection against tracing the location of a mobile subscriber by listening to the signalling exchanges on the radio path.

- Physical security means are provided to preclude the possibility to obtain sufficient information to impersonate or duplicate a subscriber in a GSM/GPRS PLMN (Public Land Mobile Network), in particular by deriving sensitive information from the Mobile Station (MS) equipment.

- Making the user information exchanged on traffic channels not available or disclosed to unauthorised individuals, entities or processes ensures the user data confidentiality. Encryption is normally applied to all voice and non-voice communications. When necessary, the MS signals to the network indicating which of up to seven ciphering algorithms specified and required in [ETS-95a] it supports. The network does not provide services to an MS that indicates that it does not support any of the ciphering algorithm(s) required by [ETS-95a]. For the activation of the ciphering during the establishment of a call, the trigger point is, at the latest, at call initiation. In the case of handover, the trigger point is, at the latest, at the completion of handover.

- The connectionless user data confidentiality feature is ensured by not making available or disclosed to unauthorised individuals, entities or processes the user information that is transferred in a connectionless packet mode.

- The following signalling information elements related to the user are protected whenever used after connection establishment:

    International Mobile Equipment Identity (IMEI),

    IMSI,

    Calling subscriber directory number (mobile terminating calls),

    Called subscriber directory number (mobile originated calls).

The IMEI cannot be changed after the Mobile Equipment (ME) final production process and the IMSI is stored securely within the SIM (Subscriber Identity Module).

4.3.2.3   UMTS

In third-generation (3G) mobile communication systems the security has been improved in order to provide a security level comparable with that of fixed networks. The 3G security requirements have been established from the main requirements resulting from a weaknesses and threats study [ETS-00a].

As a result of an analysis performed by 3GPP (3rd Generation Partnership Project), second-generation (2G) security elements are to be retained and, in some cases, improved. The main security requirements coming from 2G and reused in the UMTS are highlighted below:

- authentication of subscribers for service access;

- radio interface encryption: the strength of the encryption is greater than that used in 2G systems;

- subscriber identity confidentiality on the radio interface;

- the SIM as a removable, hardware security module that is manageable by network operators and independent of the terminal as regards its security functionality;

- SIM application toolkit security features providing a secure application layer channel between the SIM and a home network server;

- the operation of security features is independent of the user, i.e. the user does not have to do anything for the security features to be in operation.

4.3.3   Upper Layer Internet (IP Level) Functional Security Requirements

User Terminal (GMBS user and IP-based TE) mobility on a wireless transmission medium should not create an environment that is less secure than any other part of the Internet.

In order to meet the requirement defined above, the F-ISP network implements security services that are listed in the following.

- *Authentication and authorisation*: an evolutionary approach is envisaged for the security services to be implemented in the F-ISP to support authentication and authorisation:

    step 1: a standard authentication mechanism based on the pair username and password can be considered sufficient;

    step 2: a further evolution of the authentication method in the direction of Public Key Infrastructure (PKI) is envisaged. A PKI consists of protocols, services and standards supporting public-key cryptography applications (a PKI is a preliminary step for the introduction of IPSEC);

    step 3: a Certificate Hierarchy, in which all the Certification Authorities (CAs) of the F-ISPs (one for each ISP) trust the same root CA placed in the core network, shall be implemented. The access to system resources is allowed by means of access lists (ACLs).

- *Firewall (packet filtering)*: a firewall is a computer that sits between an internal network and the rest of the network and filters incoming packets, according to various config-urable criteria. Two firewall typologies are considered:

    filter firewall: this is the simplest form of firewall which selectively allows only address filtering. This kind of firewall is very simple, is extremely easy to manage and does not introduce computational overhead on the IP packet flow.

application level gateway firewall: this can inspect all the details of an IP flow. This guarantees a more effective security solution, but introduces a computational overhead on the IP flow and is hard to manage.

- *IPSEC (and SSL) (integrity, authenticity, non-repudiation, privacy)*: the IPSEC security mechanisms are implemented in order to support (i) MS authentication and identification, (ii) confidentiality and (iii) integrity of communication. The IPSEC security mechanisms are integral parts of the IPv6 protocol. IPSEC is transparent to the application, hence it does not impose any change on the existing software. Every IPv6 host must support two security payloads, *Authentication Header (AH) and Encapsulated Security Payload (ESP)*, as IP layer security (IPSEC) mechanisms. In particular:

AH provides message authentication and integrity;

ESP provides message confidentiality and integrity. ESP may optionally provide authentication if appropriate algorithms are used.

AH or ESP does not provide non-repudiation when used with the default algorithms (keyed MD5 and DES (Data Encryption Standard)). The adoption of specific algorithms, e.g. RSA with appropriate transformation, allows the provision of non-repudiation at the IP level. However an alternative choice is to implement non-repudiation at the application layer.

The use of the AH or ESP imposes sender/receiver agreement on a key, on authentication and encryption algorithms, and on a set of additional parameters needed for the adopted security algorithms. This agreement forms a *security association* between sender and receiver. Security policy controls the choice of appropriate security association. This approach allows the easy addition of new algorithms, taking into account possible future developments.

A mobile host is authenticated to the host subnet it connects to and to the corresponding node. To prevent intrusion, the mobile host must always be registered and authenticated at the host subnet before being allowed to enter the network. An authentication procedure is also used when communicating a change in the host subnet to the corresponding node and to the home subnet. This is fundamental in preventing hostile nodes from impersonating other mobile nodes by message replication.

The Internet standard key management protocol and existing PKI is a prerequisite for wide-spread use of the IP security architecture among different organisations and persons in the F-ISP.

Every host and router in the F-ISP is provided with a digital certificate to be exhibited every time it connects to a new host subnet. IPSEC authentication of IP headers will directly provide the authentication of the binding updates. Binding updates are optional headers communicating to the home subnet and to the corresponding node a new association between mobile host and host subnet.

The use of AH and ESP increases the IP processing costs and communication latency. In the case of AH, the calculation and comparison of the authentication data cause the increased latency. In the case of ESP, the encryption and decryption of the ESP cause the increased latency. A trade-off between level of security and efficiency is considered for every class of end-users and ISP.

Although IPSEC can authenticate machines, network interfaces, routers and switches, it cannot authenticate individual users and processes on a host. IPSEC security features at the IP level can be possibly completed and integrated at the higher levels using SSL. Unlike

SSL, IPSEC is completely transparent to the application, hence it does not impose any change on the existing software.

- *Auditing*: each router and firewall in the F-ISP is responsible for logging sensitive events on the basis of an agreed log policy. Each CA also logs the sensitive events and manages the revocation list.

ACKNOWLEDGEMENT

The authors wish to thank Andrea Vitaletti (Università di Roma "La Sapienza"/Etnoteam Spa) for his valuable contribution to this work.

REFERENCES

[ETS-00a] ETSI TS 121 133 V3.1.0 (2000-01) Universal Mobile Telecommunications System (UMTS); 3G Security; Security Threats and Requirements (3G TS 21.133 version 3.1.0 Release 1999).

[ETS-00b] ETSI TR 101 329-2 V1.1.1 (2000-07) Telecommunications and Internet Protocol Harmonization Over Networks (TIPHON); End to End Quality of Service in TIPHON Systems; Part 2: Definition of Quality of Service (QoS) Classes.

[ETS-01] ETSI TS 123107 v4.1.0 (2001-06) Universal Mobile Telecommunications System (UMTS); QoS Concept and Architecture (3GPP TS 23.107 version 4.1.0 Release 4).

[ETS-02a] ETSI TS 101 329-5 V1.1.2 (2002-01) Telecommunications and Internet Protocol Harmonization Over Networks (TIPHON) Release 3; Technology Compliance Specification; Part 5: Quality of Service (QoS) Measurement Methodologies.

[ETS-02b] ETSI TS 101 329-2 V2.1.3 (2002-01) Telecommunications and Internet Protocol Harmonization Over Networks (TIPHON) Release 3; End-to-end Quality of Service in TIPHON Systems; Part 2: Definition of Speech Quality of Service (QoS) Classes.

[ETS-94] ETSI TR 003 ed. 2 (1994-10), Network Aspects (NA); General Aspects of Quality of Service (QoS) and Network Performance (NP).

[ETS-95a] ETSI GTS 02.07 v3.4.1 (1995-01) European Digital Cellular Telecommunications System (Phase 1); Mobile Stations (MS) Features (GSM 02.07).

[ETS-99] ETSI TR 101 300 V2.1.1 Telecommunications and Internet Protocol Harmonization Over Networks (TIPHON) (1999-10); Description of Technical Issues.

[IET-02] J. Rosenburg *et al.*: SIP: Session Initiation Protocol, *Internet Engineering Task Force*, RFC 3261, June 2002.

[IET-94] W. Simpson (Ed.): The Point-to-Point Protocol, *Internet Engineering Task Force*, RFC 1661, July 1994.

[IET-96] H. Schulzrinne, S. Casner, R. Frederick, V. Jacobson: RTP: A Transport Protocol for Real-Time Applications, *Internet Engineering Task Force*, RFC 1889, January 1996.

[ITU-00a] ITU-T Recommendation I.356, B-ISDN ATM Layer Cell Transfer Performance (03/2000).

[ITU-00b] ITU-T Recommendation G.107, The E-Model, a Computational Model for use in Transmission Planning (05/2000).

[ITU-00c] ITU-T Recommendation P.310, Transmission Characteristics for Telephone Band (300 – 3400 Hz) Digital Telephones) (05/2000).

[ITU-00d] ITU-T Recommendation H.323, Packet Based Multimedia Communications Systems (11/2000).

[ITU-88] CCITT Recommendation G.711, Pulse Code Modulation (PCM) of Voice Frequencies (11/88).

[ITU-90] CCITT Recommendation G.726, 40, 32, 24, 16 kbit/s Adaptive Differential Pulse Code Modulation (ADPCM) 1990.

[ITU-93a] ITU-T Recommendation I.350, General Aspects of Quality of Service and Network Performance in Digital Networks, Including ISDNs (03/93).

[ITU-93b] ITU-T Recommendation I.352, Network Performance Objectives for Connection Processing Delays in a ISDN (03/93).

[ITU-94] ITU-T Recommendation E.800, Terms and definitions related to the Quality of Service and Network Performance including Dependability (08/94).

[ITU-96] ITU-T Recommendation T.120, Data Protocols for Multimedia Conferencing (07/96).

[ITU-99a] ITU-T Recommendation E.721, Network Grade of Service Parameters and Target Values for Circuit-Switched Services in the Evolving ISDN (05/99).

[ITU-99b] ITU-T Recommendation E.721, Quality of Service, Network Management and Traffic Engineering—Traffic Engineering—ISDN Traffic Engineering (05/99).

[ITU-99c] ITU-T Recommendation G.109, Definitions of Categories of Speech Transmission Quality (09/99).

ACRONYMS

2G	Second-Generation
3G	Third-Generation
3GPP	3rd Generation Partnership Project
ABR	Available Bit Rate
ACL	Access List
AH	Authentication Header
BE	Best Effort
BTS	Base Transceiver System
CA	Certification Authority
CAC	Call Admission Control
CBR	Constant Bit Rate
CC	Connection Control
CM	Connection Management
CoA	Care-of-Address
DES	Data Encryption Standard
DNS	Domain Name System
E2E	End-to-End
EFR	Enhanced Full Rate
ESP	Encapsulated Security Payload
ETSI	European Telecommunication Standards Institute
F-ISP	Federated ISP
FTP	File Transfer Protocol
FR	Full Rate
GEO	Geostationary
GMBS	Global Mobile Broadband System
GMMT	GMBS Multi-Mode Terminal
GPRS	General Packet Radio Service
GSM	Global System for Mobile Communications
HTTP	Hypertext Transfer Protocol

IETF	Internet Engineering Task Force
ILS	Internet Locator Service
IMEI	International Mobile Equipment Identity
IMSI	International Mobile Subscriber Identity
IMTC	International Multimedia Telecommunication Consortium
ITU	International Telecommunication Union
IP	Internet Protocol
IPSEC	IP Security
ISHO	Inter-segment Handover
ISP	Internet Service Provider
IWF	Inter-Working Function
LDAP	Lightweight Directory Access Protocol
M-ESW	Mobile EuroSkyWay
ME	Mobile Equipment
MCHO	Mobile Controlled Handover
MIME	Multipurpose Internet Mail Extension
MS	Mobile Station
NRT	Non Real-Time
OBP	On-Board Processing
PDCP	Packet Data Convergence Protocol
PDP	Packet Data Protocol
PKI	Public Key Infrastructure
PLMN	Public Land Mobile Network
PPP	Point to Point Protocol
PSTN	Public Switched Telephone Network
QCIF	Quarter Common Interface Format
QoS	Quality of Service
RM	Resource Management
RSVP	Resource Reservation Protocol
RTP	Real-time Transport Protocol
RTCP	Real-time Control Protocol
SIM	Subscriber Identity Module
SIP	Session Initiation Protocol
SMTP	Simple Mail Transfer Protocol
SSL	Secure Sockets Layer
TIPHON	Telecommunications and Internet Protocol Harmonization over Networks
TMSI	Temporary Mobile Subscriber Identity
TRM	Traffic Resource Manager
UDP	User Datagram Protocol
UMTS	Universal Mobile Telecommunications System
UPC	Usage Parameter Control
URL	Uniform Resource Locator
VBR	Variable Bit Rate
VoIP	Voice over IP

5

End-to-End Quality of Service Support

GIUSEPPE BIANCHI[1], NICOLA BLEFARI-MELAZZI[2],
FRANCESCO DELLI PRISCOLI[3],
PAOLO DINI[3] and MAURO FEMMINELLA[4]

[1]*University of Palermo, Italy;* [2]*University of Rome "Tor Vergata", Italy;* [3]*University of Rome
"La Sapienza", Italy; and* [4]*University of Perugia, Italy*

5.1 INTRODUCTION

5.1.1. Background

The classical framework for Quality of Service (QoS) control in packet networks is mainly based on call level control schemes [XIA-99, WHI-97]. They aim at avoiding congestion through Connection Admission Control (CAC) functions regulating the admittance or denial of a connection set-up on the basis of a set of user-declared Traffic Descriptors (TDs) at the call set-up phase. If the connection is accepted, then by monitoring the information stream (Usage Parameter Control (UPC)), the conformity of the traffic to the parameters negotiated at connection set-up can be checked. The CAC requires the network operator to be capable of estimating the performance of the whole network.

This framework has been proposed both for the ATM/B-ISDN (Asynchronous Transfer Mode/Broadband Integrated Services Digital Network) and for the Integrated Services (IntServ) Architecture (ISA) of the Internet. In fact, even if the above functions are performed differently in these two paradigms and different signalling protocols are used (Q.2931 plus Network to Network Interface (NNI) signalling in the B-ISDN, RSVP (Resource Reservation Protocol) in IntServ), the conceptual way of operation is the same and takes the form of a "hard guarantee" between the user and the network. This guarantee is one in which, if the ingress traffic conforms to a certain profile, the egress traffic maintains that profile state, and the network does not distort the desired characteristics of the end-to-end (E2E) traffic expected by the requestor. To provide such hard guarantees, the network must maintain a state (transitive for RSVP, hard for ATM) along a determined path, where each router allocates resources in compliance with the negotiated traffic profile and passes this commitment along to a neighbouring router that is closer to the nominated destination

Space/Terrestrial Mobile Networks. Edited by R.E. Sheriff, Y.F. Hu, G. Losquadro, P. Conforto, C. Tocci
© 2004 John Wiley & Sons, Ltd ISBN: 0-470-85031-0

and also capable of committing to honour the same traffic profile. This is done on a hop-by-hop basis along the selected path from the sender to the receiver (and again from receiver back to sender in RSVP). This type of state maintenance is viable within small-scale networks, but at the heart of large-scale public networks, such as the global Internet, the cost of state maintenance becomes overwhelming.

In fact, there is now a broad consensus that such QoS structures are excessively complex: ATM by itself would not be the choice for supporting QoS services because of the complexity involved, and the ISA together with RSVP is also excessively complex and possesses poor scaling properties. The main concerns about wide-scale deployment of RSVP lie in its scalability and in the resource requirements (computational processing and memory consumption) for implementing it on routers, which increase in direct proportion to the number of allocated RSVP sessions. Therefore, supporting a large number of RSVP reservations could introduce a significantly negative impact on router performance.

These scaling concerns suggest that organisations with large, high-speed networks will be reluctant to deploy RSVP in the foreseeable future, at least until these concerns are addressed.

The alternative to state maintenance and resource reservation schemes is the use of mechanisms for preferential allocation of resources, based on the deployment of different levels of Best Effort (BE). The absence of E2E guarantees of traffic flows removes the need of maintaining per-flow state information, so that "better-than-best-effort" traffic with classes of distinction can be constructed inside larger networks. Currently, the most promising direction for such "better-than-best-effort" systems appears to lie within the area of modifying the network layer queuing and discarding algorithms. These mechanisms rely on an attribute value within the IP packet's header, so that queuing and discarding decisions can be made at each intermediate node. The Internet Service Providers' (ISPs') routers must be configured to handle packets based on their IP (Internet Protocol) precedence level, or similar semantics expressed by the bit values defined in the IP packet header.

There are several methods that have been proposed within the Internet Engineering Task Force (IETF), which may yield robust mechanisms and semantics for providing these types of "differentiated services" (DiffServ).

The generic DiffServ deployment environment is based on the assumption that the network uses ingress traffic policing, where traffic entering the network is passed through traffic shaping profile mechanisms, which bound average and peak rates, and set the relative packet priority and discard criteria in accordance with the traffic profile and the administrative agreement with the customer. These ingress filters can be configured either to discard out-of-profile packets or to mark them with the lowest priority so that they are carried within the network only when there are adequate resources available. Within the core of the network, Weighted Fair Queuing (or similar proportional scheduling algorithms) can be used to allocate network resources according to the marked priority levels, allowing the high-speed and high-volume-switching component of the network to operate without maintaining per-flow state information. It is worthwhile observing that in this framework the routers/switches do not have to exchange explicit signalling messages to maintain states and to operate in a coordinated way. On the contrary, each router has only to apply a specific Per Hop Behaviour (PHB), which defines how the router must handle each received packet. Each PHB is executed locally in an independent way.

The cumulative behaviour of such stateless, local-context and distributed algorithms and corresponding deployment architectures can yield the capability of supporting distinguished and predictable service levels and holds the promise of excellent scalability. In this way, it is possible to deliver differentiated levels of performance in a manner that is predictable, fairly consistent, while providing discriminated service levels to different customers and to different applications.

It is stressed that this approach does not provide hard performance guarantees to the users. The users' perceived performance depends on the overall network dimensioning, which is an off-line and not a real-time procedure, and on the network congestion along the data path at session time, which cannot be exactly predicted.

5.1.2   Definitions

In the following sections, the notations below will be adopted:

1. A sub-network is a part of the overall Internet that is able to operate autonomously by using its own network layer protocols; when the TCP/IP (Transmission Control Protocol/ Internet Protocol) suite is put on top of the sub-network protocols and suitable inter-working devices interconnect various sub-networks, the users of the interconnected sub-networks can exchange information between each other. From this point of view, for instance, the telephone network, an Ethernet Local Area Network (LAN) and the access segments of the Global Mobile Broadband System (GMBS), that is Mobile EuroSkyWay (M-ESW), General Packet Radio Service (GPRS), Universal Mobile Telecommunications System (UMTS) and the Wireless Local Area Network (W-LAN), are sub-networks;

2. The term "IntServ approach" provides performance guarantees, by means of a three-phase procedure (TD Declaration, CAC and UPC). This approach may exploit the RSVP, other ad-hoc protocols or even may rely directly on the resource management schemes of underlying sub-networks, especially those based on connection-oriented transfer modes.

3. The term "DiffServ" approach, provides preferential treatment to specific classes of traffic but does not provide performance guarantees, does not require a state maintenance and executes pre-defined PHBs.

Under point 3 above, the detailed deployment of the DiffServ approach can follow two alternative solutions.

The first one, "Pure DiffServ", assumes that the entire network operates according to this paradigm. The only operation additional to the PHBs is the filtering and policing of the ingress traffic.

The second one, "Hybrid IntServ-DiffServ" (or "Edge and Core") approach, assumes that some sub-networks adopt IntServ while others adopt DiffServ. Typically, the sub-networks operating with IntServ are the access networks, while the core part of the network is based on DiffServ. However, the IntServ approach could be adopted also in core sub-networks with limited capacity, while some access networks could adopt the DiffServ approach. It is clear that in the Hybrid IntServ-DiffServ approach, the operations performed in each sub-network must be suitably harmonised with those relevant to other sub-networks, to provide E2E QoS levels.

To this end, suitable agreements are stipulated among the involved sub-networks. Such an agreement is called a Service Level Agreement (SLA) (see Chapter 1). The way adopted by a given sub-network to forward traffic on behalf of other sub-networks clearly determines the performance perceived on an E2E basis.

An SLA may include traffic conditioning rules that specify traffic classification rules and any corresponding traffic profiles and metering, marking, discarding and/or shaping rules, which are to be applied to the classified traffic streams.

Furthermore, an SLA can be static or dynamic. Static SLAs are the norm at present. They are agreed after negotiation between human agents and can be periodically renegotiated (days... weeks...). An SLA can specify a time varying service (times of day).

Dynamic SLAs may change frequently, as a result of variations in offered traffic load or from changes in pricing offered by the provider. Dynamic SLAs change without human intervention and require an automated agent and protocol (e.g. "Bandwidth Broker").

It is now assumed, without loss of generality, that the edge of the network adopts IntServ and the core relies on DiffServ and that the involved sub-networks comprise of the following devices, as shown in Figure 5.1:

- Leaf Router (LR);

- Edge Router (ER);

- Core Router (CR);

- Bandwidth Broker (BB): a resource manager that controls all hosts and routers of a given sub-network.

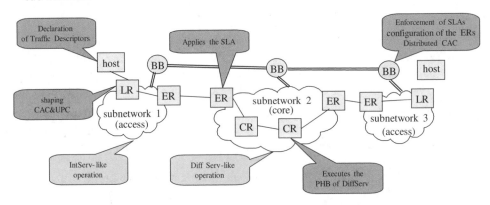

Figure 5.1   Edge-core approach of DiffServ.

When a user intends to transfer traffic with a desired QoS, they request the service to a LR, communicating a set of parameters that describe the traffic that will be generated (TDs, TD Declaration). The LR exerts a CAC and eventually accepts the requested connection. During the information transfer, the LR also executes a UPC, by monitoring the information stream, to ensure that the user emits traffic in conformity with the traffic characteristics specified at connection set-up. In this way, the major complexity is located at the edge of the network.

The CRs and the ERs implement the DiffServ mechanism by means of suitable PHBs (or scheduling discipline).

The ERs have also the specific task of guaranteeing the respect of the agreed TDs and user perceived performance between different sub-networks (SLAs). The BB exchanges signalling information with peer BBs and configures the ERs accordingly to achieve such a goal.

Finally, note that it is necessary to specify an SLA:

- At the interface between user and network;

- At the interface between any pair of communicating domains in case different DiffServ domains make up the DiffServ sub-network, each one owned and managed by a different operator.

This means that the definition and enforcement of SLAs is a requirement even in the pure DiffServ approach. Figure 5.2 shows this concept: sub-networks A and B could be two IntServ sub-networks or an IntServ sub-network and a DiffServ one, or two domains of a DiffServ region.

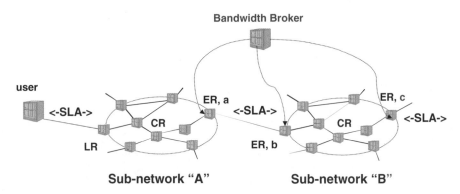

Figure 5.2 SLAs between different network entities.

5.2 QoS-RELATED GMBS ARCHITECTURE

5.2.1 General Considerations

One of the most challenging objectives of the GMBS is to support Internet QoS-sensitive services providing the users with *E2E QoS guarantees*.

Considering that the IntServ and the DiffServ architectures can be seen as complementary technologies in the pursuit of IP E2E QoS, the solution adopted in the GMBS is based on a hybrid IntServ-DiffServ approach. The rationale is to exploit on one side the possibility for the hosts to request quantifiable resources along E2E data paths, possibly provided by the IntServ architecture, and on the other side the scalability properties provided by the DiffServ architecture.

According to the hybrid IntServ-DiffServ scenario, four sets of activities need to be performed:

- IntServ operation;

- DiffServ operation;

- Coordination of the resource management among different sub-networks and DiffServ domains and interface between edge and core (definition and enforcement of SLAs);

- Fair and adaptive distribution of resources among competing sources.

In the GMBS, an E2E communication path (e.g. from a Mobile Terminal (MT) to a Wired Host) typically includes two different domains: the wireless segments domain, including the M-ESW, GPRS, UMTS and W-LAN networks, and the terrestrial Internet network domain represented by the Federated ISP (F-ISP) network.

In order to provide IP E2E QoS guarantees in each domain, appropriate and specific "QoS procedures" have to be executed, and a suitable co-ordination between the operations performed in the wireless segments and those executed in the terrestrial Internet network has to be ensured.

The QoS over a complex system, such as the GMBS, is achieved through the establishment of a QoS contract, on an E2E basis. The contract specifies all the basic characteristics that are to be guaranteed to the new connection once it has been accepted. Therefore, the problem of guaranteeing an adequate QoS to a certain connection can be conceptually divided into two parts:

- Verifying the availability of the requested resource over the whole path and, if it is the case, accepting the connection or negotiating a new contract.

- Once the connection has been activated, continuously monitoring the QoS of the connection in order to guarantee that the stipulated contract be respected.

A *QoS contract* includes the following fundamental information:

- The definition of "compliant traffic", i.e. the traffic relevant to the connection (in the IntServ approach) or to the class of connection (in the DiffServ approach) which, in any situation, has to be admitted into the network; the agreed QoS performance must be guaranteed for this kind of traffic.

- The definition of the QoS performance of the compliant traffic. The QoS performance is usually expressed in terms of maximum transfer delay, maximum loss and maximum delay jitter undergone by the Protocol Data Units relevant to the connection (or to the class of connections).

- The definition of a rule for handling the "non-compliant traffic"; possible rules could be to discard this kind of traffic, to downgrade it as a BE traffic, or to admit it into the network guaranteeing the QoS performance only if the network can afford it.

- A QoS contract has an E2E meaning.

When an access network is selected to support information transfer between an MT and the terrestrial Internet network with a required QoS, the bandwidth assignment within the wireless segment should be executed taking into account the status of resources occupancy of the core network. This resource management co-ordination is required not only at connection set-up but also while the connection is in progress. The rationale is to avoid redundantly allocating access segment resources whenever the core network is not able to ensure the requested QoS (due to current load or congestion status). This is mainly required for the M-ESW system whose resources could be more critical than those of the terrestrial wireless systems.

The hybrid IntServ-DiffServ approach, conceptually shown in Figure 5.3, foresees that all the wireless segments, acting as access networks, provide QoS guarantees by adopting the

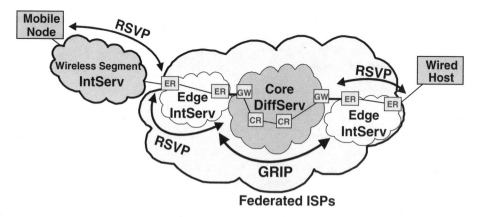

Figure 5.3   GMBS hybrid IntServ-DiffServ approach for IP QoS support. © IEEE 2002.

IntServ solution, while as far as the F-ISP network is concerned, it is envisaged that some sub-networks adopt the IntServ approach and others the DiffServ solution. More precisely, the sub-networks at the edge of the F-ISP network implement the RSVP signalling protocol by means of which strict QoS guarantees can be provided, whereas in the core of the F-ISP network, where scalability is a stringent requirement, the scalable DiffServ model is adopted.

Moreover, as the DiffServ approach provides distinguished and "predictable" service levels ("better than BE" traffic) but does not provide strict E2E QoS guarantees, Gauge&Gate Reservation with Independent Probing (GRIP) is implemented in the core portion of the F-ISP network in order to improve the user-perceived QoS performance. The GRIP mechanism combines an admission control operation, driven by end-points, with run-time traffic measurements, performed within each router to detect congestion. The detailed operation of GRIP is discussed later in the Chapter.

5.2.2   Fully IntServ End-to-End Path

It is important to note that the proposed hybrid IntServ-DiffServ Internet architecture, by means of the edge sub-networks implementing the RSVP, provides the opportunity to establish a fully IntServ-compatible E2E communication path (e.g. from an MT to a Wired Host) by-passing the F-ISP core network and thus providing per-flow hard QoS performance.

The M-ESW segment acting as the access network to the terrestrial Internet network and implementing smart routing functionality can achieve a fully IntServ-compatible E2E communication path. Such functionality allows the intelligent selection, on the basis of the Wired Host IP address, of the nearest Fixed Earth Station (FES) to the Wired Host. For example, consider the case of mobile originated communications between a satellite MT and a Wired Host. When the M-ESW access segment is selected as the supporting network, smart routing/selective landing functionality would allow the selection of the FES directly connected to the ER of the edge sub-network the wired host is connected to. Hence, the terrestrial path is minimised and it takes place only within the edge sub-network implementing the RSVP. It is evident that this will be viable every time the terrestrial edge sub-network the wired host is connected to is equipped with at least a satellite access point, i.e. a FES connected to a router of the sub-network.

5.2.3   Design Considerations

The delivery of E2E QoS using the RSVP over the wireless access networks and over the DiffServ core portion of the F-ISP network requires two main issues to be investigated:

1. The IntServ operation over the DiffServ network and the relevant service mapping;

2. The mapping between the IntServ/RSVP requests and the underlying capabilities of the GMBS access segments.

As far as the first issue is concerned, specific *Gateways* (GWs) are introduced at the interface between the DiffServ region and the IntServ region of the F-ISP network.

For what concerns the second issue, each wireless segment, selected to support the data transfer between the Sender application and the Receiver application, has to inter-work with the RSVP and activate specific QoS support mechanisms.

To this end, the approach adopted in the GMBS envisages the introduction of a QoS Support Module (QASM) implemented in the Terminal Inter-Working Unit (T-IWU) of the GMBS Multi-Mode Terminal (GMMT) and, at the network side, in the Inter-Working Unit of the FESs and in the Gateway GPRS Support Nodes (GGSNs).

The basic function of the QASM is to perform inter-working with the RSVP and, consequently, to trigger segment specific admission control procedures to verify if the currently active wireless network can accept a new connection while guaranteeing the respect of the desired QoS.

Moreover, in order to harmonise the degree of QoS offered by the different wireless segments, the QASM provides additional QoS support functionality, to those wireless networks without this functionality.

A detailed description of the QASM architecture and its basic functionality now follows.

5.3 QoS MANAGEMENT OVER THE ACCESS NETWORKS

5.3.1 Introduction

Each access network has its own specific mechanisms to manage user QoS requirements. One of the main challenges of GMBS inter-working is to harmonise these different ways of operating so that the whole system can be seen, from a user perspective, as a single network providing global coverage and QoS-guaranteed multimedia services. The QASM has been designed in order to provide users with the same QoS regardless of the access segment supporting the communication. This module is placed, at the terminal side, in the Terminal Equipment (TE), at the network side, in the GGSNs for the GPRS/UMTS network and in the FESs, for the M-ESW network. From a layering point of view, the QASM layer is placed over the access network specific layers (i.e. the radio-technology dependent layers specific to the various access networks, also referred to as Underlying Network Layers (UNLs)) and below the IP layer. A key issue of the QASM layer is that it is transparent with respect to both IP and the access network specific layers, i.e. all QASM functionalities do not have any impact on the way of working of either the IP layer, or the access network specific layers. This means that no QASM-specific header is added to the IP datagrams crossing the QASM.

The aim of the QASM is to perform functionalities (such as classification, scheduling, traffic shaping and so on) with the goals of helping to respect the QoS requirements and of enhancing the efficiency of bandwidth exploitation. Each of the above-mentioned functionalities could be either enabled or disabled depending on the involved access network; for instance, a functionality x could be enabled for all IP flows directed to the UMTS network and disabled for all IP flows directed to the GPRS network. The criterion for enabling/disabling the various functionalities is as follows: a given functionality x is enabled for IP flows directed to the access network y if and only if the functionality x in the access network y is either absent, or working less efficiently than the corresponding functionality in the QASM. This means that the QASM standardises the QoS offered to the various access networks.

As noted, an IntServ approach has been adopted in the access networks, which allows a tighter tracking of QoS connection requirements. In particular, the RSVP signalling protocol has been considered [IET-94, IET-97]. In this respect, the QASM intercepts the RSVP signalling in a read-only way, i.e. the QASM reads the RSVP information, but according to the above-mentioned transparency paradigm, does not perform any modification of this information. The reading of the RSVP information allows the QASM to be aware of the QoS connection parameters and hence to take appropriate QoS support actions, as explained in the following. In light of the above, it should be evident that the QASM must be inserted within RSVP-aware nodes; in this respect, TEs, GGSNs and FESs are appropriate hosts.

The QASM layers placed at different network entities communicate through specific QASM signalling. In this respect, each QASM is able to generate QASM Signalling Datagrams (QSDs), which are forwarded to the Underlying Networks (UNs) as if they were standard IP datagrams (i.e. from the UN point of view the QSDs look like standard IP datagrams). The QSDs are intercepted by the receiving QASM and are not forwarded to the upper layers.

In addition to RSVP signalling, basic QASM inputs are the measurements performed on the UNLs. For instance, the present status of access network specific resources (e.g. bandwidth) is key information for many QASM procedures.

5.3.2   Quality of Service Support Module

5.3.2.1   *Functionality*

The QASM implements the following QoS related functions:

- Interfacing the wireless access networks with the overall GMBS network from the QoS point of view. In this respect, the QASM is able to communicate with the RSVP Intermediate Entity (RSVPIE)[1].

- Performing a set of QoS related support functions such as traffic shaping, congestion control, scheduling, CAC and so on. These actions aim at improving the QoS in the access network that has been selected as most appropriate for handling the connection.

- Monitoring the access network status. The monitored parameters are used for assisting QoS related support functions.

The QASM functional architecture is shown in Figure 5.4. Essentially, the QASM consists of two major sub-blocks: the QoS Packets Processing Handler and the QoS Inter-working Handler.

The functional entities (FEs) belonging to the QoS Packets Processing Handler sub-block processes all IP datagrams crossing the QASM. These FEs are:

- The *Congestion Control Module* (CCM), which is in charge of shaping, policing and queuing IP traffic;

- The *Scheduler Module*, which is in charge of assigning the right priority to IP packets;

- The *Congestion Measurement Handler*, which is in charge of setting the CCM according to the information stored in the Measurement Database.

The QoS Inter-working Handler deals with the handling of RSVP signalling, UN information and QASM specific signalling. The QoS Inter-working Handler consists of the following FEs:

- The *Interceptor & Generator of QASM* (IGQASM) specific signalling, which is in charge of intercepting and generating the QSDs;

- The *RSVPIE Handler* which, during each RSVP connection set-up phase, is in charge of interfacing with the RSVPIE, in order to make the QASM aware of the RSVP parameters agreed for the new connection;

---

[1]This is the wireless segment RSVP entity participating in the E2E RSVP signalling process. It is in charge of intercepting and reconstructing the RSVP messages in order to extract RSVP parameters, which are passed to the QASM (and, in particular, as shown later, to the RSVPIE Handler) [TOC-01].

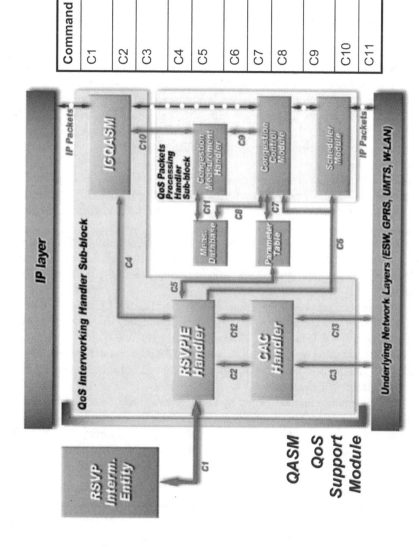

Command	Description
C1	RSVP parameters exchange between RSVPIH and RSVPIE
C2	RSVP connection set-up request response
C3	Segment Specific Connection set-up request and response
C4	QASM packet generation command
C5	Admitted connection registration/deregistration action
C6	QASM configuration command
C7	Parameter Table consultation
C8	Measurement Database consultation
C9	Congestion Control measurement and control
C10	Measurement Database updating
C11	Packet generation command

Figure 5.4 QASM functional model and internal signalling.

● The *CAC Handler*, which is in charge of triggering the CAC procedures of the UNs.

The QASM functional model also includes two databases, namely the QoS Parameter Table and the Measurement Database. The QASM FEs use these databases in order to store useful information on the QoS requirements of the in-progress connection (static information) and on the UN status (dynamic information), respectively.

5.3.2.2   QoS Packet Processing Handler

In order to manage QoS, the various connections are grouped into QoS classes. A given QoS class includes all in progress connections characterised by the same QoS requirements (e.g. in terms of bandwidth, transfer delay, jitter, loss).

The QoS Packet Processing Handler, using different approaches tailored to the specific QoS class requirements, processes every class.

In order to group connections into classes, a key factor is the connection delay sensitivity (e.g. video conferencing applications are extremely delay sensitive, whilst e-mail applications are delay insensitive). Moreover, a suitable correspondence between the QASM QoS classes and the UN QoS classes (e.g. UMTS, GPRS and M-ESW QoS classes) should be foreseen.

A large number of classes allow a better fitting between the connections and their QoS requirements, but this also increases the complexity of the QoS Packet Processing Handler. As a result of this trade-off, four classes have been chosen, as summarised in Table 5.1.

When a new connection is set-up, a new data flow (both at terminal side and at network side) crosses the QASM and the CCM, which together with the Scheduler Module, is in charge of providing the appropriate QoS support. Figure 5.5 shows the internal structure of these modules.

The QoS Switcher processes the IP datagrams entering the CCM, and is in charge of sorting the IP datagrams according to the QoS class they belong to. The sorted IP datagrams are sent to a battery of Dual Leaky Buckets (DLBs) [ELW-97], which are in charge of performing traffic shaping and traffic policing. The IP datagrams exiting the DLBs are sent to a battery of QoS queues. The IP datagrams are retrieved from these queues following the decision of a QoS Scheduler.

The DLBs' parameters are configured by taking into account both the static RSVP *Tspec* parameters, which are stored in the QoS Parameters Table at connection set-up, and dynamic information about access network status stored in the Measurement Database. If the compliant traffic denotes the traffic that complies with the DLB parameters, while the

Table 5.1   QASM QoS classes and relevant QoS constraints

QASM QoS classes	Typical applications	QoS constraints
1st Class	Voice over IP, Video conference	Very low transfer delay, Delay jitter
2nd Class	Real-Time Audio-Video (MPEGs)	Low transfer delay, Delay jitter
3rd Class	Web browsing, Telnet	Packet loss, Round-trip delay
4th Class	SMS, e-mail, FTP	Packet loss

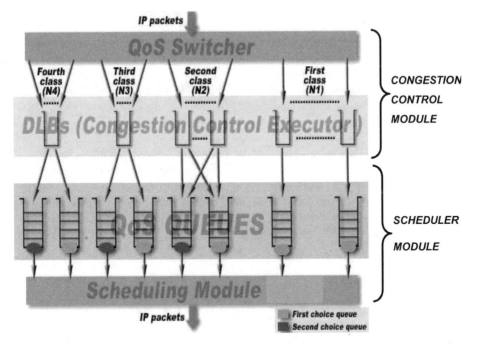

Figure 5.5   Congestion control and scheduler module.

non-compliant traffic denotes the traffic exceeding the QoS parameters, then second-choice queues are foreseen for the latter.

The Scheduler Module takes the decision on IP datagram retrieval based on a priority assigned to each datagram; this priority depends on the QoS class, the queue the IP datagram belongs to and on the arrival time of the IP datagram at the QoS queues.

In order to manage the four different QASM QoS classes, for both traffic shaping/policing and QoS queuing, a hybrid IntServ-DiffServ approach is adopted by using IntServ and/or DiffServ for each QASM QoS class, as described in the following.

5.3.2.3   QASM QoS Classes

First QASM QoS Class

Connections belonging to this class have very stringent delay jitter and transfer delay requirements; on the other hand, a certain datagram loss can be accepted. This class requires fast packet processing. This is obtained by processing the IP datagrams belonging to this class in a completely IntServ way, which means that each connection has its own DLB and QoS queue.

The reservation of one queue for each flow ensures a very low delay jitter. Queue length is then dimensioned only considering the maximum transfer delay allowed by the connection.

In this QoS class, the non-compliant traffic is discarded, because of the high transfer delay requirements.

Second QASM QoS Class

All connections belonging to this class have more relaxed requirements on transfer delay than the first QoS class connections, but stringent delay jitter requirements.

An IntServ treatment is then assumed for the compliant traffic. In view of the more relaxed transfer delay requirements, a second-choice queue is adopted for storing the non-compliant traffic. This means that the traffic stored in this queue is assigned a lower priority than the first-choice compliant traffic.

In order to reduce complexity, only one QoS queue for all second-choice class traffic is adopted. In doing so, the second-choice IP datagrams are delayed, but this is not critical for this class.

The maximum delay jitter allowed by the application dimensions the second-choice QoS queue. The worst case (i.e. maximum delay jitter) occurs when the jth IP datagram belonging to a second-choice connection enters the queue and finds the buffer empty; then the $(j + 1)$th packet, belonging to the same connection, enters the queue when the buffer is full. In this case, the relative delay between the two packets is proportional to the buffer length.

Third and Fourth QASM QoS Classes

Third and fourth QASM QoS classes are less delay sensitive than first and second classes, but have stringent loss requirements. They are managed according to a DiffServ approach: for each of these two QoS classes only one DLB and one QoS queue are present; for both of these QoS classes, the non-compliant traffic is sent to a second-choice queue.

The only difference between these two QoS classes concerns the way in which the Scheduler Module handles them: the third QoS class has the higher priority.

Scheduler Module

The selected approach for the Scheduler is the Earliest Deadline First [DEL-99]. Each QoS queue is associated to the maximum delay tolerated by the IP datagrams stored in the queue (the maximum tolerated delay for a certain connection is deduced at connection set-up as part of the RSVP-to-UN QoS mapping procedure described in the following). Then, each IP datagram stored in a QoS queue x is associated a Time-to-Live equal to the difference between the maximum tolerated delay associated to the QoS queue and the time already spent by the IP datagram in the QoS queue. Within the IP datagrams stored in the first-choice queues (the same applies within the second-choice queues, in case no IP datagram is present in any first-choice queue), the IP datagram selected for transmission is always the one with the lowest Time-to-Live.

5.3.2.4   QoS Inter-working Handler

The QoS-related GMBS architecture foresees that each edge sub-network implementing the RSVP corresponds to an ISP domain of the F-ISP network, whereas different ISP domains compose the core portion of the F-ISP network.

The term IntServ/RSVP refers to a prevailing model of RSVP usage, which includes RSVP signalling with IntServ parameters, IntServ admission control and per-flow traffic control at network elements.

According to the proposed hybrid IntServ-DiffServ approach, an RSVP session start-up takes place in the MT or Wired Host according to whether the MT or Wired Host is the Sender of application data. All the RSVP messages exchanged between the MT and the Wired Host during an RSVP session evolve transparently on the core portion of the F-ISP network where, as the RSVP is designed to operate correctly through the non-RSVP cloud, the non-RSVP capable routers do not affect these messages. In other words, these messages will be "ignored" by the DiffServ routers and as the routing of RSVP messages is independent of the reservation mechanism itself, they will be routed through the core portion of the F-ISP network exactly as if it supported RSVP.

RSVP

The RSVP treats data flow as simplex, i.e. it requests resources in only one direction and logically distinguishes the role of data Sender from that of data Receiver, although the same application process may act as both Sender and Receiver at the same time. This implies that conversational applications require two RSVP unidirectional flows that have to be individually specified.

The data Receiver originates the resource reservations (RSVP is receiver-oriented) and the state set in the routers is periodically refreshed ("soft" state) by RSVP control messages sent by the end points in order to cope with dynamic membership changes and dynamic changes of routing paths.

It is worth highlighting that the "soft" state is an RSVP key feature that well fits to the GMBS E2E QoS requirements, especially when an Inter-segment Handover/IP level Handover is in progress and routing changes along the E2E communication path are enforced.

The main RSVP message types are the Path message and the Resv message:

- Each Sender host transmits Path messages downstream towards the Receiver along the uni/multicast routes provided by the routing protocol following the path of the data. These messages set the "path state" in each node along the path. This path state includes at least the previous node unicast IP address which will be used to route the Resv messages in the reverse direction.

- Each Receiver host sends RSVP Resv (reservation request) messages upstream towards the Sender. These messages, which specify the desired QoS, follow exactly the reverse path the data packets will use during the effective data transfer phase. They create and maintain the reservation state in each node along the path and finally are delivered to the Sender.

- Path messages are sent with the same source and destination addresses as the data, so that they will be routed correctly through non-RSVP clouds. On the other hand, Resv messages are sent hop-by-hop; each RSVP-speaking node forwards a Resv message to the unicast address of a previous RSVP hop. A Path message contains the following information:

Session, which includes the destination IP address, port and protocol Identifier;

Phop, which is the address of the previous RSVP node;

Sender Template, which contains the IP address and port of the Sender. This parameter has the same expressive power and format as *Filterspec* that appears in the Resv message;

Sender_Tspec, which defines the traffic characteristics of the data flow that the Sender will generate;

Adspec, which is originated by the sending application and may be modified and updated by subsequent network elements as the Path message moves from Sender to Receiver. The *Adspec* carries information needed by the Receiver to choose a service and determine the reservation parameters. It includes both parameters describing the properties of the data path (whether or not there is a non-RSVP hop along the path; whether or not a specific QoS control service, e.g. Guaranteed Service (GS), is implemented at every hop along the path, properties of the data path itself irrespective of the selected QoS control service) and parameters required by a specific QoS control service to operate correctly.

A Resv message includes the following information:

Session, which includes the destination IP address, port and protocol Identifier;

Reservation style, which can be Fixed Filter Style, Shared Explicit Style, Wildcard Filter Style;

Filterspec, which defines the set of data packets ("flow") that has to receive the QoS defined by the *Flowspec* (see next point). The *Filterspec* contains source IP address and port and is used to set the parameters in the packet classifier of each RSVP node along the path;

Flowspec, which defines the reservation request and is used to set the parameters in the packet scheduler of each RSVP node along the path. The *Flowspec* includes the requested service class (GS or Controlled Load Service (CLS)) and two sets of numeric parameters: a *Tspec* that describes the data flow and a *Rspec*, which defines the desired QoS when a GS is requested.

The *Tspec* parameters are:

Token Bucket Rate or Sustainable Rate [r];

Peak Data Rate [p];

Token Bucket Size [b];

Minimum Policed Unit [m];

Maximum Packet Size [M].

The *Tspec* parameters [r], [p] and [b], which correspond to DLB parameters (see next section), reflect the traffic parameters of the reservation desired by the Receiver.
The *Rspec* parameters are:

the Rate [R] selected to obtain bandwidth and delay guarantees;

the optional Slack Term [S], which represents the difference between the desired delay and the delay obtained by using a reservation level R. This parameter, when specified, can be used by a network element to reduce its resource reservation for the relevant flow.

When an RSVP router receives a Path message it: creates or updates a path state associated to the corresponding session, containing *Phop*, *Sender_Tspec* and *Adspec*, restarts the cleanup timer associated to this path state (*refresh*), updates *Phop* and *Adspec* objects and then forwards the Path message towards the next router.

At each RSVP node along the reverse path the *Flowspec* parameters carried by the *Resv* message are submitted to the Admission Control test for acceptability and then used to set the scheduling process. The Admission Control test determines if the node has sufficient resources available to meet the reservation request. If the Admission Control is successful, the RSVP node forwards the *Resv* message to the previous node of the path (*Phop*); if this is not the case, an RSVP error message (*ResvErr*) is sent to the Receiver that has originated the request.

Example Operation

In the following, in order to describe the QoS Inter-working Handler and to give a preliminary definition of the GMBS E2E QoS procedures, the case of mobile-originated unidirectional communication is considered. In particular, it is assumed that the MT wishes to establish a communication with a Wired Host connected to an edge sub-network of the F-ISP network (see Figure 5.3). From the RSVP way of operation perspective, this means that the IP-based TE of the GMMT is the Sender of the application data and the Wired Host is the Receiver. For the sake of simplicity, it is assumed that only one segment of the GMBS is active and that the user cannot be simultaneously connected to more than one access network (the case of multiple active segments is only an extension of this case).

In the selected wireless segment, in order to support the RSVP Path message, a connection has to be established by activating a Connection Set-Up procedure.

More precisely, according to whether the selected wireless segment is the M-ESW segment or the GPRS/UMTS segment, a M-ESW Connection Set-Up procedure or a PDP (Packet Data Protocol) Context Activation procedure is triggered.

When a user triggers the Connection Set-Up procedure, the Sender application formulates the RSVP Path message containing the *Sender_Tspec*, describing the traffic the application expects to generate, and also constructs an initial *Adspec* object. The corresponding IP datagram is forwarded towards the access network that is currently active.

It is important to note that the *Connection Profile* of the established M-ESW connection, as well as the *QoS profile* associated with the created GPRS/UMTS PDP context, cannot refer to the desired QoS which is requested by the Receiver until after it receives the Path message. Therefore, when an RSVP Path message has to be transmitted from the MT to the Wired Host, the M-ESW connection or GPRS/UMTS PDP context is created referring to a BE service class.

As soon as the M-ESW Connection Set-Up procedure or PDP Context Activation procedure has been successfully completed, the Path Message generated by the Sender is carried over the wireless segment. Then, the GGSN (or the FES) is responsible for the standard RSVP Path processing and forwarding the Path message towards the Receiver via the external IP backbone.

When the Receiver receives the Path message, the data in the *Sender_Tspec* object and *Adspec* object are passed across the RSVP Application Program Interface to the application, which interprets the arriving data and uses it to generate the *Flowspec* parameters. These parameters are included in the RSVP Resv message, which is transmitted upstream towards the Sender.

In the DiffServ core portion where the GRIP protocol is implemented, admission control is performed, the detailed operation of which is discussed in Section 5.4.2.

After a certain time, according to the standard RSVP procedure, the GGSN (or the FES) receives from the IP backbone an RSVP Resv message containing the information concerning the resources to be reserved for the connection that is to be set-up. Then, the RSVP Intermediate Entity (RSVPIE) in the GGSN (or the FES) is responsible for intercepting the RSVP Resv message and extracting the *Tspec*, *Rspec* and *Flowspec* values (see [IET-94]), i.e. the RSVP QoS profile. This information is passed to the RSVPIE Handler, which checks the actual resource availability in the selected UN through the Call Admission Control Handler (CACH).

At first, the RSVPIE Handler maps the RSVP QoS profile onto a UN specific QoS profile (the *UN QoS profile*), according to the "*RSVP-to-UN QoS mapping procedure*", which will be described in the following section. Afterwards, the UN QoS profile is passed to the CACH, which interacts with the UN specific CAC procedure, in order to check the availability of the resources necessary for satisfying the UN QoS profile of the incoming connection, thus eventually meeting even the RSVP QoS profile. The UN QoS profile is also used to initialise the FEs of the QoS Packets Processing Handler in the GGSN (or FES) QASM.

If the UN avails of the requested resources, then the RSVP Resv message is forwarded downstream until it reaches the MT that generated the RSVP Path message (i.e. that

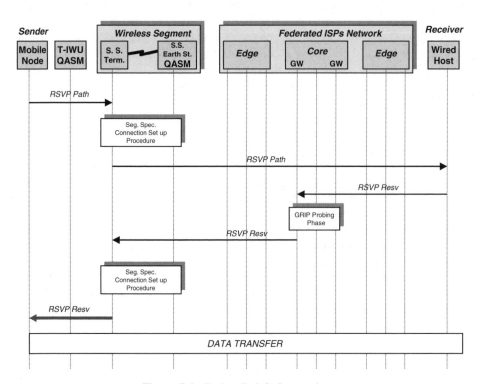

Figure 5.6 End-to-End QoS procedure.

triggered the connection set-up). This message is intercepted by the MT RSVPIE, which extracts the Resv QoS profile and passes it to the RSVPIE Handler in the MT QASM. The latter performs the RSVP-to-UN QoS mapping procedure and passes the UN QoS profile to the CACH in the MT QASM. The latter, by interacting with the UN engages the resources necessary to meet the UN QoS profile.

Figure 5.6 represents the message sequence chart diagram of the described E2E QoS procedure in the successful case.

5.3.3   RSVP-to-UN QoS Mapping Procedure

The first basic issue of this procedure is how to map the RSVP QoS Classes, namely GS, CLS and BE [IET-97] onto the four QASM QoS classes that were defined previously.

The most appropriate mapping is described in Figure 5.7. Since the QASM architecture guarantees an IntServ handling only to its first- and second-choice traffic, GS should be brought into first and second service classes; at the same time, since BE traffic is mapped onto the fourth class, then the third class remains for CLS.

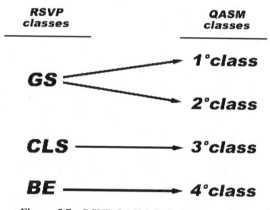

Figure 5.7   RSVP-QASM QoS class mapping.

The second basic issue of the RSVP-to-UN QoS mapping procedure is the mapping of RSVP parameters onto QASM parameters. First of all, since the QASM architecture foresees a set of DLBs for the traffic shaping/policing, it is necessary to set up their parameters according to the *Tspec* of the incoming flow. For this purpose, the RSVP *Tspec* parameters are "translated" into the relevant DLB parameters in the following way:

[Tspec] Sustainable Rate = [DLB] Token Bucket Rate;

[Tspec] Peak Data Rate = [DLB] Peak Data Rate;

[Tspec] Token Bucket Depth = [DLB] Token Bucket Size.

Then, the RSVPIE Handler based on the *Rspec* parameters received from the RSVPIE deduces the maximum tolerated delay for the connection that is to be set-up.

All these parameters are passed to the QoS Packets Processing Handler, which are used to initialise its FEs (DLBs, QoS queues and so on).

In order to define an appropriate mapping between the IntServ/RSVP requests and the underlying capabilities of the GMBS wireless segments, as well as the IntServ mapping onto the DiffServ network, it has been noted above that the current IntServ architecture provides three Service Classes: BE, CLS and GS.

The CLS is intended to support a broad class of applications that are highly sensitive to overload conditions. In particular, this service class is considered for adaptive real-time applications with relaxed delay requirements. The QoS offered by a CLS is similar to that achievable by a BE service in an unloaded network. In other words, the CLS does not accept or make use of specific target values for control parameters, such as delay or loss. Instead, acceptance of a request for CLS is defined to imply a commitment by the network element to provide the requestor with a service closely equivalent to that provided to uncontrolled (BE) traffic under lightly loaded conditions. Consequently, applications that require a CLS specify only the *Tspec* parameters, the *Rspec* parameters are not required; non-conforming traffic is treated as BE.

On the other hand, GS is considered for "real-time applications" with tight delay requirements. The QoS provided by a GS implies: assured level of bandwidth (guaranteed throughput), mathematically bounded E2E delay (guaranteed maximum delay) and no queuing losses for conforming packets. The applications that require a GS specify both the traffic characteristics (*Tspec*) and reservation characteristics (*Rspec*); non-conforming traffic is treated as BE.

When a GS class is selected by the Receiver application, each network element involved in the E2E data path, (from the Sender to the Receiver), in order to support the requested service, has to execute an admission control test to evaluate if it has sufficient resources to reserve the requested rate [R].

In the GMBS, this is required not only at each RSVP-node but also in the selected wireless segment and in the DiffServ core portion, which can be seen as network elements in the total E2E path.

As far as the wireless segment is concerned, this implies on one side that each GMBS segment has to be characterised by error terms C^2 and D^3 and on the other side that a request of GS with rate [R] for a certain flow has to be accepted only if the wireless segment can guarantee both the bandwidth [R] and no queuing loss for the conforming packets of the relevant flow.

Consequently, when a Path message is received by the QASM, the latter is in charge of updating the GS data fragment of the *Adspec* object according to the C and D error terms of the wireless segment. This allows the Receiver application to know the network path and to generate the *Rspec* parameters according to the desired E2E delay.

It is evident that the rationale of GMBS IP E2E QoS provision is that both the wireless segments and the DiffServ core portion of the F-ISP network play the role of "network elements" in the IntServ framework and are used as components of an overall E2E IntServ QoS solution. In particular, this is the approach proposed by the IETF for IntServ operation over DiffServ networks [IET-00a] and relevant mapping [IET-00b]. In order to complete the overall GMBS E2E QoS scenario, the main aspects of this approach, adopted in the GMBS,

---

[2]The error term C is the rate-dependent error term.

[3]The error term D is the rate-independent, per-element error term and represents the worst-case non-rate-based transit time variation through the service element.

as far as the access and core parts of the F-ISP network is concerned, are discussed in the remainder of this Chapter.

5.3.4 IntServ Operation over the Access Networks

5.3.4.1 Concepts

This section deals with specific aspects related to the IntServ applicability within the GMBS access segments: GPRS, UMTS and M-ESW.

As stated previously, the IntServ approach requires a traffic control structured into three phases: TD declaration, CAC and UPC. Over the last few years, countless CAC rules have been proposed in the literature and different sets of TDs have been assumed (e.g. in the ATM community), but a fully all-encompassing solution has yet to be found. One of the reasons for this is *the lack of a simple and standardised source traffic model*. In fact, the existence of many types of sources (e.g. voice, MPEG, FTP traffic, Web traffic, signalling traffic, etc.) implies the definition of source models and policing algorithms for each of them. The task of defining a single feasible and simple CAC for all of them would be too complex. This implies that in principle each source requires its own set of TDs and policing algorithms.

To solve this problem, it has been proposed to adopt a standardised regulator, the DLB, which regulates the traffic emitted by each source before entering the network. In this way, the traffic that is offered to the network has known characteristics, independent of the specific kind of source that loads the DLB. The DLB can therefore be considered as a standard interface between users and the network. The DLB has been standardised by the ATM Forum and ITU-T, and has also been proposed for the Internet.

The adoption of DLBs, together with suitable queuing disciplines and allocation rules, has allowed the definition of the actual standard IETF IntServ model, based on the RSVP signalling protocol and on the GS and CLS classes.

However, such devices can be applied also in specific traffic control frameworks, without necessarily relying on the queuing disciplines and signalling protocols adopted in the standard IntServ model. In other words, due to these traffic regulators, it is possible to design architectures that allow the provision of QoS guarantees. One such example is the standard IntServ model, using the already defined service classes, but other alternatives are possible. Such alternatives include:

1. The definition of additional service classes in the RSVP framework;

2. The definition of service provisioning in autonomous frameworks, not necessarily based on the same way of operation as RSVP;

3. The definition of allocation rules, complemented by suitable management and signalling procedures that allow integrating in a "larger" and standard RSVP environment, specific infrastructures that cannot/do not want to apply in each network node the per-flow queuing discipline and allocation rules of the two standard RSVP service classes. In other words, a specific segment can provide QoS guarantees to the traffic that it handles by means of "local" procedures; this means that when such a segment is integrated in an E2E path that crosses also other segments and sub-networks (operating standard RSVP procedures) it does not endanger E2E performance;

In the following, general issues that are common to all DLB-based traffic control architectures are first discussed and then these general principles are applied to the GMBS access segments, by focusing on alternative (3) above.

A DLB implements jointly two Generic Cell Rate Algorithms (GCRA) [ATM-96, ITU-96]. These GCRAs ensure that the traffic entering the system can be characterised by only four parameters: the Peak Cell Rate (PCR) and its Tolerance, and the Sustainable Cell Rate (SCR) and its Tolerance (or Burst Tolerance). From the parameters of the GCRA, it is easy to obtain those characterising the DLB. DLBs can be used to smooth the incoming traffic in order to make its statistical profile more compliant with the characteristics of the underlying bearer service. This approach greatly simplifies the three phases of a traffic control scheme. In fact, the three traffic control phases particularise in:

1. *Choice of the TDs*: the user must simply choose the values of four parameters so as to achieve a compromise (ideally optimal) between the desired QoS and the cost of the requested resources. For instance, a low value of the SCR could degrade performance but it could reduce the price for the user; a value of the SCR near the PCR could increase the costs charged by the network operator but it would speed up communications. The four TDs relevant to a given application could be either chosen once and for all, or configured by the users dynamically, according to their needs. Once a traffic source is suitably characterised, it is not difficult to choose TDs in such a way as to obtain the desired performance (see Figure 5.8).

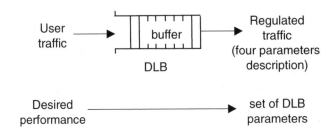

Figure 5.8 DLB operation.

2. *CAC procedures*: the network must define CAC rules and dimensioning criteria based on the above TDs. All the traffic entering the network is characterised only by means of the above TDs. It was stated previously that countless CAC rules have been proposed in the literature. Due to the significant simplification implied by the use of only four TDs, the problem becomes simpler. A first solution is the one relevant to the standard RSVP model and to the actually defined service classes. Such a solution requires suitable per-flow queuing discipline and allocation rules at each network node and guarantees a bounded delay and no loss phenomena. Another solution is that proposed in [ELW-97] and adapted to the M-ESW system in [BLE-99]. Such a solution is focused on the definition of allocation rules that allow guaranteeing pre-defined and arbitrary delay and loss values, chosen by the network operator. In other words, the latter proposal is concerned only with allocation rules; such rules can be used to define additional service classes in the RSVP framework, or to design traffic control schemes in specific

environments; in the latter case, the allocation rules must be complemented by suitable management and signalling procedures.

3. *Policing*: the DLB regulator simply executes the policing functions. In fact, the defined TDs have been designed to be very easily policed.

In the following, it will be assumed that all the QoS-sensitive traffic entering the GMBS is regulated by means of DLB regulators.

Therefore, the system will have the following characteristics:

1. A clear and "standard" interface between application and network layers can be designed so that even different protocols can operate over the network layer by simply calling suitable procedures and different network layers can be seen in the same way by the application layer.

2. By means of the above procedures, the upper layers can request specific services to the network layer, with pre-defined performance and QoS guarantees, or simply on a BE basis.

3. This leads to a modular and flexible system, where different services and paradigms can be simultaneously or alternatively used, by simply changing part of the software.

4. The key to this approach is the interface between application and network/transport layers (or between user and network) and the "standardised" way used by the upper layer modules (or by the users) for requesting a specific service with a specific QoS. The latter interface will be based on DLB regulators. The adoption of such regulators allows also the coexistence of the above-described framework with reactive congestion control schemes such as the Available Bit Rate (ABR) traffic category, defined by the ATM Forum and for the support of IP BE traffic [BAI-99, BLE-99].

It should be noted that the adoption of DLBs implies the above advantages but has also the disadvantage of the relevant cost. However, a regulator is needed even in the pure DiffServ approach and that, in any case, such a device is needed for charging purposes. In fact, the operator must measure the user traffic in order to charge it and the DLB can be certainly used for this purpose. This means that what has been defined as a cost for the proposed solution is in reality something that must be provided in any case. With this consideration it may then be deduced that the presented solution has no extra cost, so leaving only the advantages.

In the following, examples of how the CAC concepts presented above can be applied in the GMBS access segments are presented.

5.3.4.2 GPRS and UMTS

The way of operation is in principle compatible with the DLB operation. It is in fact possible to map DLB parameters to throughput, peak and mean bit rates, transfer delay and priority. However, to complete the standard IntServ/RSVP model it would be required to implement in each network node suitable queuing disciplines and allocation rules. To avoid such modifications of the GPRS segment, alternative (3) described in the previous section is selected: the definition of allocation rules, complemented by suitable management and signalling procedures that allow integrating in a "larger" and standard RSVP environment, the GPRS segment. The mapping of DLB parameters onto segment specific characteristics and the relevant way of operation is of pertinence to the QASM, as discussed earlier in the Chapter.

The same philosophy can be applied to the UMTS segment. In conclusion, suitable techniques allow integrating the GPRS and the UMTS segments into a larger standard RSVP environment.

5.3.4.3   M-ESW Segment

This case is more complex and requires some preliminary considerations.

Initially, an original traffic control scheme that is specific to the M-ESW segment and that is logically located below the IP level is presented. Then, how the M-ESW segment can be integrated into a larger standard RSVP environment is discussed.

It is well known that packet switching provides flexibility and efficiency [e.g. BAI-96, BAI-99, TOH-98]. However, it appears that full packet-switches are still considered complex and expensive, when implemented on-board satellites (at least in the mid-term). In addition, circuit-switching has still some attractions, at least for particular services. For the time being, a compromise between circuit and packet transfer modes has been found in satellite networks with Dynamic Bandwidth Allocation Capabilities (DBAC). Such systems are based on classical circuit-switches, but the DBAC payload allows changing dynamically the bandwidth of each connection, without the need of tearing-down and setting-up the connection itself. The M-ESW network is a DBAC system.

As has been noted in previous Chapters, the M-ESW network is composed of:

- A certain number of GEO satellites;

- Network Operation Centre (NOC);

- User terminals (mobile);

- FESs, including a certain number of Inter-working Units (IWU); these IWUs are stations that interface the M-ESW system with terrestrial core networks (e.g. the Internet).

The traffic control methodology is IntServ-like. The set-up procedure is handled by the NOC; therefore the signalling information has to make two satellite hops (lasting two satellite round trip times) to go from a terminal to the NOC and back. Once a connection is set-up, the DBAC allows the changing of the bandwidth of the connection itself by means of In Band Requests (IBRs). Each user can request to increase or decrease dynamically (frame by frame) the capacity of a connection, according to their needs. The IBRs are sent to a module resident in the satellite, called Traffic Resource Manager (TRM). The TRM can accept or reject the IBRs. When a user requests a change of capacity, the new value of capacity is effective only after the reception of a positive answer from the TRM. The latency of the proposed scheme is then equal to one satellite round trip time, ΔT. Due to the DBAC, M-ESW remains a fairly simple system (the on-board switch is a simple T-switch) and yet it can offer increased efficiency and flexibility.

To simplify the system procedures, two assumptions are made:

1. The overall traffic is divided into high priority (HP) traffic and low priority (LP) traffic. A given amount of resources is assigned to the HP traffic so as to satisfy pre-defined

performance. When the HP traffic is not exploiting all the allocated capacity, the unused capacity is temporarily taken away from the HP traffic and used to carry LP traffic. When the HP traffic needs the capacity that has been previously taken away, this capacity is once again given back to the HP traffic itself, after a round trip time. These variations in the use of the capacity by the HP traffic are performed by means of the DBAC.

2. The HP traffic entering the system is regulated by means of DLBs.

In the following, a generic satellite terminal (this can be an MT, a FES or an IWU) is considered and it is assumed that this terminal intends to generate a given amount of traffic towards the satellite network. It is also assumed that each terminal has a buffer dedicated to the HP traffic (HP buffer) and that, if various traffic sources belonging to the same terminal want to emit traffic towards the same destination, then they can be statistically multiplexed over the same M-ESW connection. This can be an unusual case for an MT, but it can be a frequent situation for IWUs or for connections carrying traffic on behalf of Service Providers.

The architecture of a satellite terminal (SaT) is shown in Figure 5.9. For the sake of simplicity, only one M-ESW connection is considered. The HP traffic generated by a SaT and supported by the same M-ESW connection can be a superposition of K hetero-geneous traffic sources. An additional buffer (LP buffer) can be eventually used by the LP

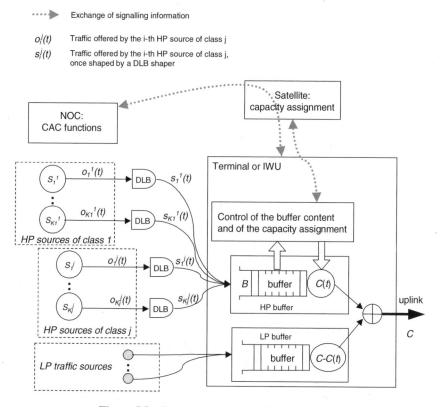

Figure 5.9 System architecture of a generic SaT.

traffic. The resource management scheme must ensure that the performance measures perceived by the HP traffic are always those that are contracted, while no guarantee is given to the LP traffic. The HP traffic could consist of GS traffic. The LP traffic could be an IP BE.

With reference to the traffic coming out from a generic DLB and feeding the satellite network, the following are denoted as:

P_S: the Peak Cell Rate (in cells/s);

r_S: the Sustainable Cell Rate (in cells/s);

B_{TS}:   the Token Buffer Size (in cells) (i.e. a parameter related to the Tolerance of the Sustainable Cell Rate, or Burst Tolerance).

It is assumed that the Tolerance of the Peak Cell Rate is equal to zero or included in the Peak Cell Rate parameter. Due to the above assumptions, each HP source is described by means of three TDs (i.e. P_S, r_S and B_{TS}).

Two ways of operation are envisaged:

Case a: deterministic QoS guarantees; at connection set-up, the SaT requests the NOC to establish a connection with a capacity equal to C cells/s. The NOC accepts connection requests until the sum of the relevant capacities is less than or equal to the overall capacity of the involved links (i.e. with a peak allocation). Given the circuit-switching transfer mode of the M-ESW system, this means that the SaT requests a *circuit* with capacity C. However, the HP traffic uses transport resources only when they are effectively needed and the capacity actually used is then a function of time, $C(t) \leq C$. The remaining capacity $C - C(t)$ can be used by LP traffic (belonging to the same SaT or to other SaTs). The capacity C is then shared between HP and LP traffic in a dynamic way, as a consequence of IBRs. With this approach, the HP traffic does not experience loss phenomena.

Case b: statistical QoS guarantees; this operates as above but with the difference that the NOC does not exert a peak allocation but a statistical one; loss phenomena can occur, but a suitable CAC ensures that the performance measures perceived by the HP sources are always those requested.

The above solutions concern the CAC operation of the NOC (denoted by in the following as "NOC related CAC rules"). It is clear that in Case a above, these rules become that of trivial peak allocation while in Case b a statistical CAC is needed. However, the DBAC implies the necessity of another set of functions and rules, which can be considered as of pertinence to the SaT.

In particular, a methodology to choose the values of the system parameters so that the HP users perceive the desired performance is foreseen. For instance, given K sources, an HP buffer of size B and the TDs of each HP traffic source, it may be needed to determine the value of the connection capacity C such that the users' requirements in terms of loss and delay are satisfied. On the other hand, given C, B and the TDs of each HP traffic source, it may be of interest to determine the value of K such that the users are satisfied with the perceived performance, even in presence of the DBAC mechanism. The former case is generally called a dimensioning problem, while the latter is referred to as definition of CAC rules. For the sake of simplicity, in the following the use of "SaT related CAC rules" will be used, being noted that the same rules can be used both for dimensioning and for resource allocation purposes.

The DBAC has then a twofold application:

1. At NOC level, it improves the system efficiency by allowing the allocation of a number of SaTs greater than the number that would be achieved by means of a peak allocation (in Case *b* above);

2. Once a SaT is assigned a given connection, the bandwidth may be handled dynamically, without involving the NOC, so that also LP traffic can be delivered, and the channel utilisation factor improves further.

Finally, the possible statistical multiplexing of K sources onto the same M-ESW connection (see Figure 5.9) helps in increasing the overall efficiency.

To complete the definition of the scheme, how and when a SaT decides to request more capacity to transport HP traffic, or to release a portion of the capacity that has been used for the HP traffic, must be described. This is achieved by assuming that the NOC CAC procedure has assigned a given bandwidth, say C, to a connection in a given SaT. Eventual modifications of the capacity actually used may happen in every frame, but, when the SaT requests for a bandwidth variation, this variation is effective only when the TRM module on-board the satellite acknowledges the request.

The algorithm used to dynamically assign capacity to the HP traffic of a given connection (and, as a consequence, to the LP traffic) is the following. The value of $C(t)$, that is the capacity being used by the HP traffic, is chosen as a function of the content of the HP buffer. The remaining capacity, $C - C(t)$, is offered to the LP traffic. It is clear that $C(t)$ must be a non-decreasing function of the HP buffer content: when the content of such a buffer increases, the HP traffic requires more capacity and *vice-versa*.

Due to the significant simplification implied by the adoption of the DLBs, the CAC problem becomes simpler and allows the traffic control problem in the M-ESW segment to be solved. The QoS-sensitive traffic is transported by the HP traffic category of M-ESW; the definition of allocation rules, complemented by the M-ESW management and signalling procedures allow integrating in a "larger" and standard RSVP environment, the M-ESW segment. As a consequence, it is possible to proceed as in the case of the GPRS and UMTS segments.

5.4 QoS MANAGEMENT OVER THE CORE NETWORK

5.4.1 IntServ Operation over DiffServ Networks

5.4.1.1 *Reference Network Configuration*

Provision of E2E QoS control in the IntServ model is based on the concatenation of "network elements" along the data transmission path. When all of the concatenated network elements implement one of the defined IntServ "services" (GS, CLS), the resulting data transmission path will deliver a known, controlled QoS defined by the particular IntServ service in use.

The availability of DiffServ per-hop and cloud-edge behaviours, together with additional mechanisms to statically or dynamically limit the absolute level of traffic within a traffic class, allows a DiffServ network cloud to act as a network element within the IntServ framework. In other words, an appropriately designed, configured and managed DiffServ network cloud can act as one component of an overall E2E QoS controlled data path using the IntServ framework, and therefore support the delivery of IntServ QoS services.

Figure 5.10 shows the basic use of a DiffServ network cloud as an IntServ network element. The reference network includes a DiffServ region in the middle of a larger network

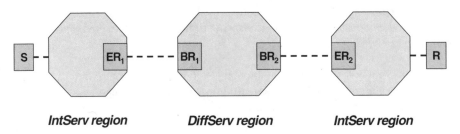

IntServ region   **DiffServ region**   **IntServ region**

Figure 5.10   Reference network configuration for IntServ operation over Diffserv networks.

supporting IntServ E2E. Both the term "DiffServ network cloud" and "DiffServ region" are used to identify a set of contiguous routers which support Behaviours Aggregate (BA) classification and traffic control.

By comparing this reference network configuration with the QoS-related GMBS architecture shown in Figure 5.3, it can be noted that the IntServ regions and the DiffServ region illustrated in Figure 5.10 correspond to the edge sub-networks and core portion of the GMBS F-ISP network, respectively; the Border Routers, BR_1 and BR_2, correspond to the GWs introduced in the F-ISP network at the interface between the DiffServ region and the IntServ region.

The required IntServ network element functions are mapped onto the DiffServ cloud as follows.

5.4.1.2   Traffic Scheduling

The IntServ traffic scheduling function is supported by appropriately selected, configured and provisioned PHBs within the DiffServ network. These PHBs, when concatenated along the path of traffic flow, must provide a scheduling result that adequately approximates the result defined by the IntServ service. In general, the PHB concatenation will only be able to approximate the defined IntServ service over a limited range of operating conditions (level of traffic, allocated resources, and the like). In that case, other elements of the network, such as shapers and policers, must ensure that the traffic conditions seen by the PHBs stay within this range.

5.4.1.3   Traffic Classification

The IntServ framework requires that each network element (re)classify arriving traffic into flows for further processing. This requirement is based on the architectural assumption that network elements should be independent, and not depend on other network elements for correct operation.

Note that the IntServ framework does not specify the granularity of a flow. IntServ is often associated with per-application or per-session E2E flows, but in fact any collection of packets that can be described by an appropriate classifier can be treated as an IntServ traffic flow.

When IntServ is mapped onto DiffServ, packets must be classified into flows, policed, shaped, and marked with the appropriate DSCP (Differentiated Services Codepoint) before they enter the interior of the DiffServ cloud. Strictly speaking, the independence requirement stated above implies that the ingress boundary router of each DiffServ cloud must implement a Multi-field classifier to perform the classification function. However, in keeping with the DiffServ model, it is permissible to push the flow classification function further towards the edge of the network if appropriate agreements are in place. For example, flows may be classified and marked by the upstream ER if the DiffServ network is prepared to trust this router.

5.4.1.4   Policing and Shaping

In terms of location in the network, these functions are similar to traffic classification. A strict interpretation of the IntServ framework would require that the ingress boundary router of the DiffServ cloud perform these functions. In practice, they may be pushed to an upstream ER if appropriate agreements are in place.

Note that moving the shaping function upstream of the DiffServ ingress boundary router may result in poorer overall QoS performance. This is because if shaping is performed at the BR, a single shaper can be applied to all of the traffic in the service class, whereas if the shaping is performed upstream, separate shapers will be applied to the traffic from each upstream node.

5.4.1.5   Admission Control

The quantitative IntServ services (GS, CLS) require that some form of admission control limits the amount of arriving traffic relative to the available resources. Two issues are of interest; the method used by the DiffServ cloud to determine whether sufficient resources are available, and the method used by the overall network to query the DiffServ cloud about this availability.

Within the cloud, the admission control mechanism is closely related to resource allocation. If some form of static resource allocation (provisioning) is used, the admission control function can be performed by any network component that is aware of this allocation, such as a suitably configured BR. If resource allocation within the network cloud is dynamic (a dynamic "bandwidth broker" or signalling protocol) then this protocol can also perform the admission control function, by refusing to admit new traffic when it determines that it cannot allocate new resources to match.

The admission control mechanism used is independent of the admission control algorithm used to determine whether sufficient resources are available to admit a new traffic flow. The algorithm used may range from simple peak-rate allocation to a complex statistical measurement-based approach.

The admission control mechanism used within the DiffServ cloud is also independent of the mechanism used by the outside world to request service from the cloud. As an example,

E2E RSVP might be used together with any form of interior admission control mechanism—static provisioning, a central BB, or aggregate RSVP internal signalling.

The key to providing absolute, quantitative QoS services within a DiffServ network is to ensure that at each hop in the network, the resources allocated to the PHBs used for these services are sufficient to handle the arriving traffic. As described above, this can be achieved through a variety of mechanisms ranging from static provisioning to dynamic per-hop signalling within the cloud. Two situations are possible:

- With per-cloud provisioning, sufficient resources are made available in the network so that traffic arriving at an ingress point can flow to any egress point without violating the PHB resource allocation requirements. In this case, admission control and traffic management decisions need not be based on destination information.

- With per-path provisioning, resources are made available in the network to ensure that the PHB resource allocation requirements will not be violated if traffic arriving at an ingress point flows to one (in the unicast case) specific egress point. This requires that admission control and resource allocation mechanisms take into account the egress point of traffic entering the network, but results in more efficient resource utilisation.

Two points are important to note:

- Both approaches are valuable, but all functions must adopt the same approach, particularly if resource allocation is per-path. In this case, traffic shaping and policing, and hence classification must be destination aware as well.

- The per-cloud versus per-path decision is independent of decisions about static versus dynamic provisioning. It is often assumed that dynamic provisioning is necessarily per-path, while static provisioning is more likely to be per-cloud. In reality, all options may be useful in differing circumstances.

5.4.2   IntServ Operation over DiffServ Networks Enhanced with GRIP

5.4.2.1   The GRIP paradigm

GRIP End-nodes Operation

GRIP's end-nodes operation is extremely simple, and is depicted in Figure 5.11 (for more details, see [BIA-02]). Consider, initially, the set-up of a mono-directional flow, i.e. from a source node to a destination node. When a user terminal requests a connection with a destination terminal, the source node transmits a *probing packet*[4]. Meanwhile, the source node activates a probing phase timer, lasting for a reasonably short time. If no response is received from the destination node before the timeout expiration, the source node enforces

---

[4]For clarity of presentation, in what follows, the source and destination user terminals are identified as the network end-nodes, taking admission control decisions. However, for obvious security reasons, such end-nodes should be the ingress and egress nodes, under the control of the network operator(s). In addition, if GRIP is seen as a Per Domain Behaviour specific mechanism, then such end-nodes are suitable nodes, located at the edge of the considered domain.

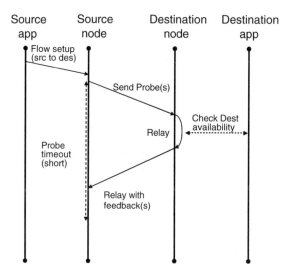

Figure 5.11   End point GRIP operation.

rejection of the connection set-up attempt. Otherwise, if a *feedback packet* is received in time, the connection is accepted, and control is given back to the user application, which starts a data phase, simply consisting of the transmission of information packets.

To maintain full compatibility with the (stateless) DiffServ architecture, probing packets are not meant to carry explicit signalling information describing the characteristics of the associated data traffic (e.g. peak bandwidth), to be interpreted by the network routers. Instead, probing and information packets are labelled with different values of the DSCP field in the IP packet header. The different tags given to probing and information packets allow internal routers to apply them to different forwarding behaviours and, as described in the following section, enforce probing packet dropping (and thus block the set-up attempt) when congestion arises.

The role of the destination node simply consists of monitoring the incoming IP packets, intercepting those labelled as probing, reading their source address, and, for each incoming probing packet, just relaying with the transmission of a feedback packet, if the destination is willing to accept the set-up request (the probing packet is not meant to carry signalling information for network routers, but it may carry application-level signalling information to be analysed at the destination node). Note that an important by-product of GRIP is that it allows a receiver capability negotiation, as requested in [IET-00a]. In other words, there is no value in the sender establishing a QoS-enabled path across a network to the receiver if the latter is incapable or simply unwilling to receive the consequent data flow. GRIP explicitly allows such feedback.

The described GRIP operation is easily extended to provide set-up for bi-directional connections. In such a case, the destination node will simply relay with a probing packet instead of a feedback packet. A feedback will be ultimately sent back by the source node upon reception of the destination probing (to close the three way connection set-up handshake—independent probing mechanisms are likely to be needed to test both "uplink" and "downlink" network paths, which may differ). Finally, GRIP can be adapted to support "downlink" (destination to source) flows. In this case, the GRIP set-up operation can be activated on the destination node by means of an application-level protocol (i.e. a video retrieval can be started via HTTP/RTSP (Hypertext Transfer Protocol/Real Time Streaming Protocol) [IET-96, IET-98], by clicking on a web menu page).

GRIP Router Operation

Figure 5.12 illustrates the GRIP operation over a DiffServ router. Packets incoming to a router output port are dispatched to the relevant queues according to their DSCP tag.

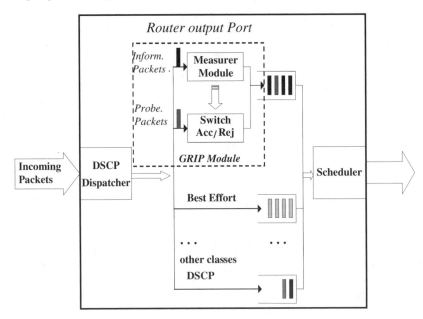

Figure 5.12   GRIP operation over a DiffServ router.

Within the router, a GRIP module is in charge of handling both probing packets and information packets. Note that several GRIP modules, devised to support different traffic classes, may coexist within the router. Recall that information packets are generated by traffic sources that have already passed an admission control test. Within each GRIP module, a *measurement module* is in charge of measuring the load offered by information packets, i.e. the overall *aggregate* accepted traffic. On the basis of these running traffic measurements, and according to a suitable *Decision Criterion*, the measurement module drives an *Accept/ Reject switch*. When the switch is in the ACCEPT state, incoming probing packets are forwarded to the output queue. Conversely, probing packets are dropped when the switch is in the REJECT state. In other words, the router acts as a gate for the probing flow, where the gate is opened or closed on the basis of the traffic estimates (hence the Gauge&Gate in the acronym GRIP).

The operation presented above, in conjunction with the end-nodes operation described in the previous section, provides a per-flow admission control function over a stateless network, via an implicit signalling pipe of which the network remains unaware. In fact, in order for a call set-up procedure to succeed, the probing packet needs to find all the routers along the path in the ACCEPT state (if the probing packet encounters a router in the REJECT state, it gets discarded; hence it does not reach the destination, no feedback packet will be relayed back, and the call will be blocked as soon as the probing phase timer expires—see also Figure 5.13 discussed below). Moreover, and most important, by leaving each router in charge of locally deciding whether it can admit new flows, GRIP allows to decouple the

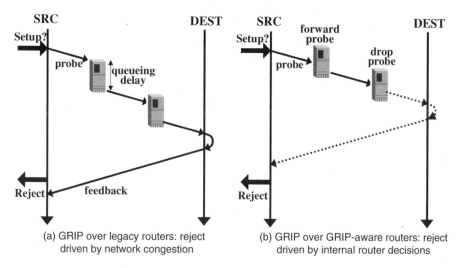

(a) GRIP over legacy routers: reject   (b) GRIP over GRIP-aware routers: reject
driven by network congestion   driven by internal router decisions

Figure 5.13   GRIP operation over different environments.

admission control function from the QoS provisioning level. Since the notion of congestion, which triggers the GRIP gate mechanism, is left to the Decision Criterion implementation, each administrative domain may select and deploy quantitative performance specifications, by suitably selecting and configuring the Decision Criterion within the domain routers.

An example of a simple Decision Criterion is to switch from ACCEPT to REJECT state when the aggregate load measurement exceeds a given threshold. The fact that a target QoS performance can be obtained by simply controlling throughput (i.e. measurements taken on accepted traffic) is well established, and is at the basis of more sophisticated state-of-the-art Measurement Based Admission Control (MBAC) implementations described in [BJS-00, GRO-99]. It is obvious that the resulting QoS performance depends upon the link capacity and the traffic model.

Finally, it is remarked that GRIP is backward compatible with existing routers. Figure 5.13 shows the case of flow rejection when GRIP is operated over a GRIP-unaware domain. In this case, flow rejection is purely driven by queuing delay caused by internal network congestion (note the difference with GRIP aware routers, which explicitly enforce probing packet dropping, see Figure 5.13). Upon congestion, the round trip delay (probing plus feedback) may become larger than the probing phase timeout, and thus a flow set-up is rejected. Stability is enforced by the fact that when network congestion increases, a corresponding decrease in the probability that set-up is successful occurs. Therefore, a lower number of new flows are set up, and this allows the network to smoothly decongest. It can be observed that this idea is close to that of the TCP congestion control technique, but it is used in the novel context of admission control. Clearly, when GRIP is operated over a GRIP-unaware network, very little can be said about the level of QoS provided.

Decision Criterion

In defining the Decision Criterion, the aim is to provide *strict QoS guarantees in all operational conditions*. To this purpose, information about QoS traffic flow characteristics is needed, to correctly reach ACCEPT/REJECT decisions. Thus, it is assumed that traffic sources are regulated at network edges by standard DLBs. This information is exploited within the measurement algorithm driving the ACCEPT/REJECT decisions, the task of which is to estimate the current load on the considered link.

In this section, only homogeneous traffic sources are considered, in the assumption that heterogeneous traffic sources are handled by separate GRIP modules. In other words, it is assumed that homogeneous flows (i.e. all flows with the same DLB parameters or *Tspec*) belong to the same traffic class and that each class is handled in a separate way and recognised by routers by assigning different *pairs* of DiffServ codepoints to different traffic classes (i.e. each class has a DSCP for probing packets and another one for information packets). Traffic sources belonging to different traffic classes are handled by separate GRIP modules and a measurement module for each class is required. In this scenario, the load estimation simply corresponds to estimating the number of active flows traversing the considered GRIP-compliant router.

It is emphasised that the assumptions are coherent with the DiffServ paradigm. In other words, the scheme is not strictly limited to only one class of traffic with the same *Tspec*. Different traffic classes are handled in a *differentiated* way.

Other ways to handle different traffic classes without using separate GRIP modules can be found in [BIA-01]. Defining the number of traffic classes and the relevant number of different measurement modules is a matter of a trade-off between complexity and efficiency. Eventually, priority schemes among traffic classes can be easily introduced. Thus, separate traffic classes, such as those considered in this Chapter, are not necessarily a synonym of complexity and simplistic treatment, but instead of flexibility, coherently with the DiffServ approach.

In any case, explicit information about the *traffic mix composition* is not required, that is how many active flows per each regulator type are present in each node. This duty is assigned to the measuring and estimator module in each node, which operates on traffic aggregates only.

5.4.2.2   GRIP Implementation Options

GRIP acts as an admission control agent to the DiffServ network region.

This can be done according to two possible options, A and B.

In option A, the GRIP control loop is executed between the end-nodes (Sender and Receiver) and the functionality of the GRIP probe, packets is executed by means of the RSVP Path messages. In other words, the Path messages carry out both RSVP and GRIP related functions: the GRIP probe, is piggybacked onto the Path message. In particular, as far as GRIP is concerned, when the upstream GW, called GW1, receives a Path message (with the added significance of a GRIP probe, packet), it executes the mapping of the RSVP QoS request onto a DiffServ GRIP class (and marks the relevant DSCP). It also starts the GRIP Probing Phase by injecting into the DiffServ sub-network the probe, (= Path) packet relevant to the selected DiffServ class. If such a packet succeeds in reaching the egress GW, called GW2, then it means that all the involved DiffServ/GRIP routers are found in the ACCEPT state and that the DiffServ region can support the requested connection. The GW2 forwards the Probe (= Path) message to the next RSVP router belonging to the adjacent IntServ region, which continues the RSVP operation, until reaching the Receiver node. If the Receiver node is willing to accept the connection, it answers with a Resv message, which has also the added significance of a GRIP feedback packet. In the meantime, the Sender node has activated a probing phase timer. If the Sender receives the feedback packet, (i.e. the Resv message) before this timer expires, the Probing Phase is successfully completed; in this case, control is given back to the user application, which starts a Data Phase, simply consisting of the transmission of data packets, which will then be marked with a suitable DSCP by the upstream GW. This operation is reported in Figure 5.14 as "Option A".

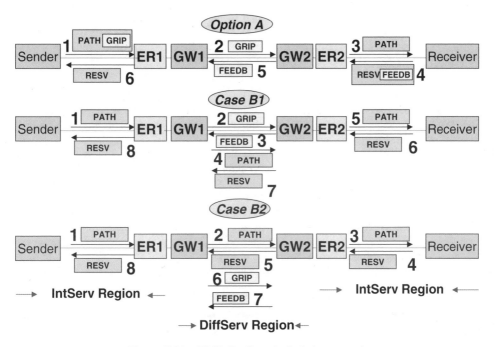

Figure 5.14   GRIP distributed admission control.

In option B, the GRIP control loop is executed between the GWs. In other words, these routers assume the role of GRIP source and destination. The mapping between RSVP requests and DiffServ GRIP classes is executed as above. The probing procedure can be carried out:

1. By storing the Path message in GW1, starting a GRIP operation between GW1 and GW2 and letting the Path message go ahead through the DiffServ network only if the GRIP operation is successful (that is if GW1 receives a feedback packet from GW2, within the GRIP timeout). Otherwise, an RSVP error message is sent by GW1 to the Sender node that originated the Path message; this operation is reported in Figure 5.14 as "Case B1".

2. By simply forwarding the Path message through the DiffServ network, without GRIP operation; then, when the Resv message is eventually received by GW1, the latter starts a GRIP operation and stores the Resv message; if the GRIP operation is successful, GW1 forwards the Resv message upstream towards the Sender node. Otherwise, an RSVP error message is sent by GW1 to the Receiver node that originated the RSVP request. In this alternative, GW1 can also exploit the information contained in the Resv message (i.e. requested bandwidth and slack term, in addition to *Tspec*) to fine-tune the GRIP request. This operation is reported in Figure 5.14 as "Case B2".

Note that the last two alternatives have the disadvantage of increasing the set-up delay.

It is worth highlighting that in the GMBS, by means of the implementation of the GRIP protocol and under the assumption that all the traffic entering the DiffServ cloud is regulated by DLBs, strict delay bounds can be provided in the core portion of the F-ISP network.

REFERENCES

[ATM-96] ATM Forum, *Traffic Management Specification*, V4.0., April 1996.

[BAI-96] A. Baiocchi, N. Blefari-Melazzi, M. Listanti, C. Soprano: An ATM-like System Architecture for Satellite Communications including On-Board Switching, *International Journal of Satellite Communications*, **14**(5), September-October 1996; 389–412.

[BAI-99] A. Baiocchi, N. Blefari-Melazzi, M. Listanti, C. Soprano: Definition and Performance Analysis of Simple, ABR-like, Congestion Control Scheme for Satellite ATM Networks with Guaranteed Loss Performance, *IEEE Journal on Selected Areas in Communications (JSAC)*, **17**(2), February 1999; 303–313.

[BIA-01] G. Bianchi, N. Blefari-Melazzi, M. Femminella, F. Pugini: Performance Evaluation of a Measurement-based Algorithm for Distributed Admission Control in a DiffServ Framework, *2001 Tyrrhenian International Workshop*, 17–20 September, 2001, Taormina, Italy (in Lecture Notes in Computer Science, Springer-Verlag, Sergio Palazzo, Ed.)

[BIA-02] G. Bianchi, N. Blefari-Melazzi, M. Femminella: Per-Flow QoS over a Stateless Differentiated Services IP Domain, *Computer Networks—The International Journal of Computer and Telecommunications Networking*, **40**(1), 16 September 2002; 73–87.

[BJS-00] L. Breslau, S. Jamin, S. Schenker: Comments on the Performance of Measurement-Based Admission Control Algorithms, *Proceedings of IEEE Infocom 2000*, **3**, Tel-Aviv, 28–30 March 2000; 1233–1242.

[BLE-99] N. Blefari-Melazzi, G. Reali: A Resource Management Scheme for Satellite Networks with Dynamic Bandwidth Allocation Capabilities, *IEEE Multimedia Special Issue on Satellite Systems for Mobile Multimedia Services*, **6**(4), October–December 1999; 54–63.

[DEL-99] F. Delli Priscoli: Design and Implementation of a Simple and Efficient Medium Access Control for High Speed Wireless Local Area Networks, *IEEE Journal on Selected Areas in Communications (JSAC)*, **17**(11), November 1999; 2052–2064.

[ELW-97] A. Elwalid, D. Mitra: Traffic Shaping at a Network Node: Theory, Optimum Design, Admission Control, *Proceedings of IEEE Infocom 97*, Kobe, Japan, 9–11 April, 1997; 444–454.

[GRO-99] M. Grossglauser, D.N.C. Tse: A Framework for Robust Measurement-Based Admission Control, *IEEE/ACM Transactions on Networking*, **7**(3), June 1999; 293–309.

[IET-00a] G. Huston: Next Steps for the IP QoS Architecture, *Internet Engineering Task Force*, RFC 2990, November 2000.

[IET-00b] Y. Bernet: A Framework for Internet Services Operation over DiffServ Networks, *Internet Engineering Task Force*, RFC 2998, November 2000.

[IET-94] R. Braden, D. Clark, S. Shenker: Integrated Services in the Internet Architecture: An Overview, *Internet Engineering Task Force*, RFC 1633, July 1994.

[IET-96] T. Berners-Lee, R. Fielding, H. Frystyk: Hypertext Transfer Protocol—HTTP/1.0, *Internet Engineering Task Force*, RFC 1945, May 1996.

[IET-97] R. Braden (Ed.) *et al.*: Resource Reservation Protoocol (RSVP)—Version 1 Functional Specification, *Internet Engineering Task Force*, RFC 2205, September 1997.

[IET-98] H. Schulzrinne, A. Rao, R. Lanphier: Real Time Streaming Protocol (RTSP), *Internet Engineering Task Force*, RFC 2326, April 1998.

[ITU-96] ITU-T Recommendation I.371, *Traffic Control and Congestion Control in B-ISDN*, (03/2000).

[TOC-01] C. Tocci, P. Conforto, A. De Carolis, M. Femminella, F. Pugini: Solutions and Techniques for the Provision of End-to-End IP QoS in a GMBS System, *Proceedings of IST Mobile Summit*, Barcelona, Spain, 9–12 September 2001; 312–317.

[WHI-97] P.P. White: RSVP and Integrated Services in the Internet: A Tutorial, *IEEE Communications*, **35**(5), May 1997; 100–106.

[XIA-99] X. Xiao, L.M. Ni: Internet QoS: A Big Picture, *IEEE Network*, **13**(2), March-April 1999; 8–18.

ACRONYMS

ABR	Available Bit Rate
ATM	Asynchronous Transfer Mode
BA	Behaviours Aggregate
B-ISDN	Broadband Integrated Services Digital Network
BB	Bandwidth Broker
BE	Best Effort
BR	Border Router
CAC	Connection Admission Control
CACH	Call Admission Control Handler
CCM	Congestion Control Module
CLS	Controlled Load Service
CR	Core Router
DBAC	Dynamic Bandwidth Allocation Capabilities
DiffServ	Differentiated Services
DLB	Dual Leaky Bucket
DS	Differentiated Service
DSCP	Differentiated Services Codepoint
E2E	End-to-End
ER	Edge Router
FE	Functional Entity
FES	Fixed Earth Station
F-ISP	Federated-ISP
FTP	File Transfer Protocol
GCRA	Generic Cell Rate Algorithm
GGSN	Gateway GPRS Support Node
GMBS	Global Mobile Broadband System
GMMT	GMBS Multi-mode Terminal
GPRS	General Packet Radio Service
GRIP	Gauge&Gate Reservation with Independent Probing

GS   Guaranteed Service
GW   Gateway
HP   High Priority
HTTP   Hypertext Transfer Protocol
IBR   In Band Request
IntServ   Integrated Services
IETF   Internet Engineering Task Force
IGQASM   Interceptor & Generator of QASM
ISA   Integrated Services Architecture
ISP   Internet Service Provider
IWU   Inter-Working Unit
LAN   Local Area Network
LP   Low Priority
LR   Leaf Router
MBAC   Measurement Based Admission Control
M-ESW   Mobile EuroSkyWay
MPEG   Moving Picture Experts Group
MT   Mobile Terminal
NNI   Network to Network Interface
NOC   Network Operation Centre
PDP   Packet Data Protocol
PCR   Peak Cell Rate
PDP   Packet Data Protocol
PHB   Per Hop Behaviour
QASM   QoS Support Module
QoS   Quality of Service
QSD   QASM Signalling Datagram
RSVP   Reservation Protocol
RSVPIE   RSVP Intermediate Entity
RTSP   Real Time Streaming Protocol
SaT   Satellite Terminal
SCR   Sustainable Cell Rate
SLA   Service Level Agreement
T-IWU   Terminal Inter-Working Unit
TCP/IP   Transmission Control Protocol/Internet Protocol
TD   Traffic Descriptor
TE   Terminal Equipment
TRM   Traffic Resource Manager
UE   User Equipment
UMTS   Universal Mobile Telecommunications System
UN   Underlying Network
UNL   UN Layer
UPC   Usage Parameter Control
W-LAN   Wireless LAN

6

Mobility Support

PAULINE M.L. CHAN[1], PAOLO CONFORTO[2],
Y. FUN HU[1] and RAY E. SHERIFF[1]

[1]*University of Bradford, UK,* [2]*Alenia Spazio, Italy*

6.1 LOCATION MANAGEMENT

6.1.1 Background

Mobility management essentially consists of two main components: location management and handover management. Location management refers to the ability of the network to determine the current point of attachment of the mobile terminal (MT) to the network, whereas handover management refers to the ability of maintaining on-going sessions when the MT moves and changes its point of access to the network.

The location management concept was historically developed in the context of mobile networks. Unlike early fixed telecommunications networks, where the terminal was unequivocally associated with a network access point (NAP), in mobile systems the connection between a terminal and a NAP can vary dynamically according to the movements of the terminal itself. The knowledge of the location of each mobile is one of the most important features of cellular systems, which need to continuously identify a mobile's position by executing appropriate location management procedures.

However, roaming is not specific only to wireless systems. Nowadays, some fixed networks provide subscribers with a roaming service defined as "the possibility of using a telecommunications terminal at a given point of the network". The subscriber is therefore characterised by an identity and can access the network through different access points. The network is able to track the location of subscribers and to route the call towards the correct point [TAB-00].

The Global Mobile Broadband System (GMBS) aims to devise a location management scheme that is more structured than those adopted in current mobile and fixed networks. The system consists of a *multi-segment access network*—where each component adopts its own mobility management scheme—integrated with the *Internet network* (the Federated Internet Service Provider (F-ISP) network, see Chapter 1) – where Mobile IP (Internet Protocol) is

Space/Terrestrial Mobile Networks. Edited by R.E. Sheriff, Y.F. Hu, G. Losquadro, P. Conforto, C. Tocci
© 2004 John Wiley & Sons, Ltd ISBN: 0-470-85031-0

implemented to provide "roaming services". In particular, the different levels of mobility entail the existence of different levels of location management:

- *Internet Network Location Management:* aiming at identifying the point of access to the Internet network, i.e. the point where IP packets have to be routed;

- *Intra-Segment Location Management:* executed by the access segment specific entities when the GMBS Multi-Mode Terminal (GMMT) moves within the access network;

- *Inter-Segment Location Management:* executed by GMBS specific entities when the GMMT moves from one access network to another access network.

A discussion on each level of location management now follows.

6.1.2 Internet Network Location Management

In Mobile IP [IET-02, IET-03a], each mobile node is assigned a pair of addresses: one for identification, i.e. the *home IP address* defined in the address space of the home link, and one for location, i.e. the *care-of-address* (CoA), defined in the address space of the visited/foreign link.

The network uses the CoA to identify the current position of a mobile node, i.e. the current position of a given subscriber. This continuous tracking of the subscriber's location allows the Internet to provide a subscriber with roaming services.

In fixed networks where roaming services are supported, the position of subscribers should always be stored in appropriate databases. In the Internet where Mobile IP is implemented, the routing tables (also referred to as binding tables) play the role of location database in the *home agent* (HA) and in the *correspondent node*. These tables store the correspondence between the mobile node home address, i.e. the subscriber identifier, and the CoA, i.e. the location identifier. By using the information in the binding table, it is possible to correctly route the IP packets towards the Internet point of access where the mobile node is connected.

Adopting the same rationale for the GMBS, the location management in the F-ISP network is based on the main features of Mobile IP, exploiting the binding between home addresses and CoAs stored in the home agent and in the correspondent node tables.

The GMMT (or equivalently the Terminal Inter-Working Unit (T-IWU)), on the other hand, can be seen from the Internet perspective as a mobile node, or to be more precise, a mobile router. Once the GMMT selects the access segment (i.e. Mobile EuroSkyWay (M-ESW), General Packet Radio Service (GPRS)/Universal Mobile Telecommunications System (UMTS)) in charge of providing connectivity, the access point to the Internet network is also automatically defined. Such an access point is located at:

- the edge router of the terrestrial service provider when there is a direct interconnection between the access segment and terrestrial service provider; in this case, direct links are established between the M-ESW Fixed Earth Station (FES) and the edge router or between the GPRS/UMTS Gateway GPRS Support Node (GGSN) and the edge router;

- the NAP; in this case the M-ESW FES or the GPRS/UMTS GGSN are connected to the NAP and, through the NAP, to the border router of other terrestrial service providers.

If M-ESW is used to provide connectivity towards the Internet, the point of access to the F-ISP network will depend also on the FES selected after having executed smart routing functionality (see Chapter 5).

In the scenario where Terminal Equipments (TEs) (also referred to as Mobile Host (MH)) do not implement Mobile IP, binding tables (and therefore, location management function-ality) are located only in the mobile router home agent (MRHA) (and eventually in the correspondent node).

On the other hand, in the scenario where TEs do implement Mobile IP, binding tables (and therefore, location management functionality) can be found, not only in the MHRA, but also in the mobile host home agent (MHHA).

It is worth highlighting specific features of the Internet location management functions implemented in the GMBS. In the fixed Internet network, the routing protocols, on the basis of the knowledge of the CoA, route the IP packets directly to the mobile node, while in the GMBS, packets will be routed up to the access segment node connected to the F-ISP network by means of direct interconnections or at the NAP. An example of the point-of-access to the Internet for M-ESW and GPRS is shown in Figure 6.1. Assuming that the

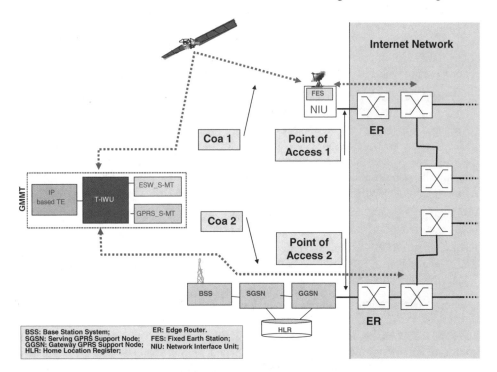

Figure 6.1   Point-of-access for M-ESW and GPRS.t141

GMMT has selected one of the GPRS/UMTS components, the Internet routing protocols route the IP packets up to the GGSN while, if the selected segment is M-ESW, IP packets are routed up to the chosen FES. Once the packets reach the GGSN or the FES, intra-segment location management procedures are activated and the routing towards the T-IWU shall be performed according to the mechanisms adopted by each access segment (intra-segment mobility). This intra-segment routing is transparent to the Internet network which, for

location management and routing purposes, needs to have only the information pertaining to
the interface towards the GGSN and the FES.

6.1.3 Intra-Segment Location Management

As stated in the previous section, on the basis of the information contained in the binding
tables, which can be considered as a type of Internet location management database, the
interface towards the network node of the GMBS access component in charge of supporting
the IP packet flow can be identified and localised. Internet protocols are in charge of routing
IP packets up to this interface. Once the packets are delivered to the GGSN or to the FES,
they are routed up to the T-IWU according the GPRS/UMTS or M-ESW specific mechanisms, respectively.

Each access segment implements its own (intra-segment) location management scheme.
Without going into too much detail, it is worth noting that the set of location information in
the GPRS/UMTS segments are stored in both the Mobile Station (MS) and in the Serving
GPRS Support Node (SGSN) and is referred to as the Mobility Management (MM) context.
Three mobility management states can be identified in the GPRS:

- The IDLE State: In the IDLE state no MM context is activated. The MS and the SGSN do
  not have any valid location or routing information for the subscriber.

- The READY State: After a successful completion of a GPRS attach procedure, the MS
  enters into the READY state characterised by the activation of a MM context both in the
  MS and SGSN. Such a MM context contains information about both the current routing
  area (analogous to the GSM (Global System for Mobile communications) location area)
  and the cell where the MS is roaming. If, due to its movement, the MS changes cells
  within the same routing area, a cell update procedure is executed. If it also changes the
  routing area, a routing area update procedure is performed.

- The STANDBY State: At the expiry of a suitably fixed time-out, the STANDBY state can
  be entered. This state is characterised by a MM context in both the MS and the SGSN.
  However, the SGSN MM context contains location information of the MS only at routing
  area level when the STANDBY state is encountered. Furthermore, paging requests are
  sent over the whole routing area where the MS is roaming.

In analogy, there are three mobility management states for a UMTS subscriber, namely:

- PMM-DETACHED: In the PMM-DETACHED state, no valid location or routing
  information for the MS is held by the MS and SGSN contexts. The MS cannot be
  reached by a 3G-SGSN, since the MS location is unknown.

- PMM-CONNECTED: The MM context enters into the PMM-CONNECTED state when
  an attach is performed. The MS location is known in the 3G-SGSN with an accuracy of a
  serving RNC (Radio Network Controller). In the PMM-CONNECTED state, the location
  of the MS is tracked by the serving RNC. When the Routing Area Identity information in
  the MM system changes, the MS performs the routing area update procedure. The MS
  location is known in the 3G-SGSN with an accuracy of a routing area and can only be
  reached by paging, e.g. for signalling.

- PMM-IDLE: The MS enters the PMM-IDLE state when its packet-switched signalling connection to the 3G-SGSN is released or broken. This release or failure is explicitly indicated by the RNC to the MS or detected by the MS (Radio Resource Control (RRC)-connection failure). Radio connection release will also happen if an URA (UTRAN (UMTS Terrestrial Radio Access Network) Registration Area) update fails because of "RRC connection not established", or when the URA update timer expires while the MS is out of coverage.

As far as the M-ESW segment is concerned, the location management information is stored in a database, the *Localisation Table*, located in the *Network Operation Centre* (NOC). This database stores the identifier of the M-ESW satellite spot-beam where the broadband satellite terminal (SaT) is roaming. This information is communicated by the SaT to the NOC during the *M-ESW Registration procedure* and used during the connection set-up phase in order to send a paging message in the correct spot-beam. This is discussed further in Chapter 7.

6.1.4 Inter-Segment Location Management

Inter-segment location management identifies the set of GMBS specific procedures for storing information of the GMBS multi-segment access network components available at a certain time, as well as information about the access segment selected by the T-IWU of the GMMT for each IP-based TE operated at a given time by a GMBS user.

On the basis of several parameters, such as radio coverage conditions, GMBS user profile, Quality of Service (QoS) perceived by the user, etc., the T-IWU in the GMMT continuously executes procedures with the objective of selecting the most suitable access segment. If the value of the above quoted parameters changes, a T-IWU at a given time can decide on whether to reselect the access segment for an IP-based TE in the GMMT.

Inter-segment location management implies that, at the terminal side, the T-IWU will be in charge of:

- keeping track of the available access segments, i.e. the access segments among which the most suitable one will be selected;

- implementing a suitable database where the access segment selected for each IP-based TE is stored, i.e. the result of the segment selection or inter-segment handover (ISHO) procedure for each IP-based TE. Note: ISHO is also commonly referred to as vertical handover.

At the network side, inter-segment location management has two dimensions: (i) at the IP level and (ii) at the GMBS Service Node level.

As far as location management at the IP level is concerned, the change in the access segment, which can take place both when the GMMT is in STANDBY mode or in ACTIVE mode, implies also a change in the point of access to the Internet network, resulting in a different route to be followed by the packets directed to the GMMT. In order to correctly route these packets over the Internet, it is necessary to have information on the access segment currently in use. To be more precise, it is necessary to have the information concerning the interface towards the access node of the access segment currently in use. From the Internet point of view, no additional procedure or database is required to store this information since the information on the point of access of the selected access segment is implicitly contained in the CoA assigned to the GMMT (i.e. the mobile router coinciding with the T-IWU) and therefore it is already stored in the binding table of the MRHA.

For location management, it is worth pointing out that the GMBS is operated by a GMBS Service Provider (GSP); the main objective of which is to provide nomadic GMBS users with connectivity to the Internet network with a guaranteed QoS irrespective of their position. The GSP controls a GMBS Service Node, which is a server belonging to the Internet of the GSP. The GMBS Service Node is in charge of executing several functions such as the authentication and registration of a GMBS user—suitably provided with a GMBS Subscriber Identity Module, which is a smart card storing GMBS user profile information—accessing the F-ISP network via one GMMT. In addition, the GMBS Service Node hosts a suitable database where the information pertaining to the access segment selected for a given GMBS user at any time is stored. To be more precise, this database stores the result of the segment selection or ISHO procedure for each IP-based TE. Such information is used for charging purposes, to assist satellite-to-terrestrial handover (in order to trigger the satellite resource release procedure so as to speed up the release of satellite resources) and to execute some specific resource management procedures, aiming both at retrieving information concerning the resource allocation and congestion status of the access segments and at forwarding such information, after a pre-elaboration, to the GMMT.

6.2 ADDRESS MANAGEMENT

6.2.1 Classes of Address

The network architecture and the mobility management scheme devised for the GMBS, imply that two classes of addresses can be identified:

- The underlying network addresses: these addresses have validity in the context of the components of the GMBS multi-segment access network they refer to. So GPRS/UMTS specific and M-ESW specific addresses belong to this class.

- IP addresses: these are IP addresses (IPv6 is envisaged for the target system even though compatibility with IPv4 will be taken into account) having a global validity.

The design of the GMBS address management considers the presence of both of these classes of address and their mutual relationships.

Different kinds of underlying network addresses are defined in the different components of the GMBS multi-segment access network for routing purposes.

6.2.2 GPRS/UMTS

As far as the GPRS/UMTS segment is concerned, the IP packets are transparently transported between the Internet network, in particular the edge router connected to the

GGSN (see Figure 6.1), and the MS (or the GMMT). In GPRS, this transparent transport is realised by means of encapsulation functionality that defines two tunnels: (i) one between the GGSN and the SGSN and (ii) another between the SGSN and the MS. In UMTS, two different encapsulation schemes are used: (i) one for the backbone network between two GSNs and between an SGSN and a RNC, and (ii) one for the UMTS RRC connection between the RNC and the MS.

In GPRS, encapsulation between the GGSN and the SGSN is executed by the GPRS Tunnelling Protocol (GTP), which encapsulates IP packets by adding an appropriate GTP header creating the GTP Packet Data Units (PDUs). Encapsulated packets are then inserted in TCP (Transmission Control Protocol) (or UDP (User Datagram Protocol)) PDUs which in turn are inserted in IP PDUs. The same applies to encapsulation of Packet Data Protocol (PDP) PDUs for UMTS, i.e. the packet domain PLMN (Public Land Mobile Network) backbone network encapsulates a PDP PDU with a GTP header, and inserts this GTP PDU in a UDP PDU that again is inserted in an IP PDU. In order to correctly tunnel these packets it is therefore necessary to unequivocally identify the GGSN and SGSN nodes involved in the tunnel and the particular tunnel, among those managed by the GGSN/SGSN. This is realised by

- assigning each GSN an IP address (either IPv4 or IPv6) for routing within the GPRS backbone network. These IP addresses belong to a private address space and therefore are not accessible from the public Internet.

- assigning the tunnel a Tunnel Identifier (TID) which is used to identify a PDP context. The TID consists of:

  an *International Mobile Subscriber Identity* (IMSI) which identifies the GPRS subscriber;

  a *Network Layer Access Point Identifier* (NLAPI) which identifies the network layer (i.e. the IP layer).

In UMTS, in the same fashion, the IP and GTP PDU headers contain the GSN addresses and Tunnel Endpoint Identifier necessary to uniquely address a GSN PDP context.

In GPRS, encapsulation between SGSN and MS is realised by the Sub-Network Dependent Convergence Protocol. SGSN and MS PDP context are uniquely addressed with

- a Temporary Logical Link Identity (TLLI): identifying the logical link between the MS and the SGSN (and therefore identifying the MS). The TLLI is derived from the Packet Temporary Mobile Subscriber Identity allocated at the moment of the GPRS attach.

- a NLAPI: as previously defined.

Encapsulation of PDP PDUs between 3G-SGSN and RNC on the Iu-interface is achieved by means of GTP header. PDP PDUs exchanged between RNC and MS on the Uu-interface are encapsulated with Packet Data Convergence Protocol [ETS-02c].

A GPRS subscriber identified by an IMSI will have one or more network layer addresses (i.e. PDP addresses) temporarily or permanently assigned. In the GMBS this PDP address is an IPv6 address. It is assigned to the MS by the GGSN during the *PDP context activation procedure*.

Three different ways to assign an IP address are envisaged in [ETS-02a]:

1. the Home Public Land Mobile Network (HPLMN) operator assigns an IP address permanently to the MS (*static IP address*);

2. the HPLMN operator assigns a PDP address to the MS when a PDP context (i.e. IP context in the GMBS) is activated (*dynamic HPLMN IP address*);

3. the Visited Public Land Mobile Network (VPLMN) operator assigns an IP address to the MS when a PDP context is activated (*dynamic VPLMN IP address*).

In these cases, the IP address assigned to the MS belongs to the address space of the mobile operator giving such an address. This holds both in the case of static and of dynamic assignment, as well as both for HPLMN and VPLMN operator assignment. Given that the GGSN is seen by the Internet network as a standard router with a given IP address, the IP address assigned to the MS has the same network prefix (Net_Id) as that of the GGSN IP address. In other words, the GGSN is seen by the Internet core network as an edge router providing the access to a sub-network: such a sub-network is represented by the GPRS network.

In [ETS-02b], another way to assign an MS an IP address is envisaged:

4. the PDN (Packet Data Network) operator assigns an IP address to the MS after the PDP context is activated (*external IP address allocation*).

In this case the IP address belongs to the address space of the PDN operator, i.e. to the address space managed by an ISP or to the address space of a sub-network the GGSN is inserted in. The PDP context in the GPRS network is therefore preliminary activated before the IP address assignment. Then the MS will activate a standard IP mechanism such as *stateful autoconfiguration* based on Dynamic Host Configuration Protocol (DHCP) [IET-03b] in order to receive an IP address. In such a situation, the GGSN is only in charge of relaying the DHCP messages towards the ISP or towards the sub-network it is connected to. Once the address is assigned, the PDP context is suitably modified in order that the involved GPRS entities know the IP address. The four address allocation schemes introduced above also apply to UMTS.

2G-GPRS release 1998 [ETS-02a] has been proposed as baseline for the GMBS. It is a widespread opinion that IP address dynamic assignment should be preferred to the static assignment. As a matter of fact, even though the former presents the disadvantage that a PDP context can be activated only by the MS and not by the network (the same disadvantages exists for UMTS)—this implies that if the GGSN receives an IP packet with a destination address not having any PDP context associated, the GGSN discards the packet—several benefits can be obtained:

- the mobile operator needs a smaller number of addresses.

- more cost-effectiveness for the user (dynamic addressing is also currently implemented in fixed Internet for private users using dial-in).

- possible routing optimisation when roaming: in case of static assignment, IP packets directed to the MS are always addressed towards the GGSN of the HPLMN regardless of whether the GPRS user is actually the HPLMN domain or in a VPLMN domain. With a dynamic assignment, IP packets are addressed towards the GGSN of the operator currently serving the MS so that long, and sometimes-redundant paths in the inter-PLMN GPRS network are avoided.

- since different GGSNs can offer different services or connections to different ISPs, dynamic address assignment guarantees more flexibility.

It is the responsibility of the GGSN to allocate and release the dynamic IP address, but there is no standard specifying the way this is realised. It is an implementation issue and several different solutions can be envisaged. A possibility (but it is not the only one) is to configure the GGSN to act as a DHCP server but other options can be considered depending on the platform used by the manufacturer.

In the GMBS, the TCP/IP-based TE in the GMMT can be seen as a mobile node that already has an IP address, i.e. the *home address*, before executing a PDP context activation. The IP (i.e. PDP) address assigned by the GGSN during this procedure represents the CoA assigned to the GMMT. This CoA needs to be stored in the binding tables of the mobile node's home agent and correspondent node.

One (or more) PDP context stored in the MS, in the SGSN and in the GGSN, is associated with this IP address (i.e. CoA). When IP packets with the CoA in the destination field reaches the GGSN they are routed through the GPRS network up to the GMMT according to the encapsulation mechanisms described above.

Moreover, the GPRS segment has to provide the outer Internet world with topology information needed to correctly implement routing protocols. To do this, the GGSN should be upgraded with Border Gateway Protocol (BGP) functionality to exchange routing protocol information with the other BGP speakers of Internet Autonomous Systems (AS).

6.2.3   M-ESW

As noted previously, two categories of addresses are classified: underlying network addresses and IP addresses. This classification applies also to the *M-ESW segment*. A M-ESW broadband satellite terminal (SaT) is identified by the following identifiers:

- M-ESW_Terminal_Id: This Id unequivocally identifies a M-ESW TE within the whole population of the M-ESW system. It can be seen as an addressing number. A M-ESW SaT can serve several different OLN (overlaying network) users, which exploit the satellite connectivity to access the corresponding OLN. In the GMBS, the OLN is represented by the Internet network so that the OLN user coincides with the IP-based TE present in the GMMT. According to the M-ESW specifications, the OLN user is assigned the following identifier:

- OLN_User_Id: This unequivocally identifies, within the OLN, the OLN user requesting M-ESW services. It is used for set-up procedures and for addressing purposes. This Id corresponds to the IP address of the IP-based TE in the GMMT.

The correspondence between the OLN_User_Id of a given OLN user and the M-ESW_Terminal_Id identifying the SaT that the OLN user is utilising at a certain time is stored in the M-ESW NOC in a suitable database called Localisation/Routing Table, along with the identifier of the spot-beam serving the SaT.

Considering a fixed-to-mobile communication (the same considerations apply also for a mobile-to-mobile communication), when a "calling" OLN (i.e. IP) user intends to communicate with a "called" OLN (IP) user in the GMMT through the satellite segment, a M-ESW connection has to be set-up between the FES and SaT in the GMMT. Without going into too much detail of the M-ESW connection procedure, the M-ESW NOC receives a M-ESW set-up message containing (among several fields) the OLN_User_Id, i.e. the IP address of the called OLN user. By looking up the Localisation/Routing Table using such a field as an entry, the NOC is able to determine the Terminal_Id of the SaT serving the called OLN user and the spot-beam where it is roaming. On the basis of this information it is able to correctly address the connection set-up request.

Similar considerations hold also for mobile-to-fixed communication even though some additional capabilities are used. In this case the Localisation/Routing Table shall be mainly used to select the most suitable FES with respect to the position of the correspondent node involved in a communication with the OLN user served by the satellite network, so as to implement *smart routing* functionality.

Each entry of the table, i.e. each destination or, equivalently, each "called" OLN user, is associated to a *FES field* indicating one possible FES to be used to support the communication and a further *metric field* indicating the weight for the route exploiting that FES.

The metric field is an indication of "how good is" the terrestrial path between the FES itself and final destination. Such a field is generated on the basis of information carried by the BGP messages exchanged between the M-ESW BGP speakers (located at the FESs) and the BGP speakers of other terrestrial ASs. Therefore, once the Localisation/Routing Table has been constructed, the implementation of smart routing functionality consists of a correct looking up of the table and of the selection of the FES associated to the "called" OLN user ID (i.e. IP destination address) with the smallest figure in the metric field.

After being set-up, a M-ESW connection is assigned an identifier which unequivocally associates a M-ESW connection within the M-ESW network. The OLN PDUs, which essentially are IP packets in the GMBS, to be exchanged between the called and calling terminals defined above, are therefore transmitted over the M-ESW connection with that identifier.

6.3   HANDOVER MANAGEMENT

Handover management comprises of connection management and QoS management. Conceptually handover is divided into three phases, as shown in Table 6.1.

Connection management includes the handover monitoring, handover initiation and handover control/execution procedures. The handover strategies and criteria will be investigated within the framework of handover monitoring and initiation. Procedures such as the transfer of control data, bearer set-up and release, bearer switching, route set-up and release, etc. will be investigated under handover execution.

For QoS management, the procedure to ensure the negotiated QoS during call set-up can still be supported or re-negotiated in the event of handover.

Table 6.1 Three Phases of handover

Functional phase	Process	Main task
Information Gathering phase	Monitoring, collecting and processing data	Radio link measurement
Handover Decision phase	Decision making process	Select a target cell or spot-beam
Handover Execution phase	Path creation and switching, completion and route optimisation	Establish a new connection

6.4 IP MOBILITY SUPPORT ISSUES

6.4.1 Overview

For the GMBS, two levels of mobility are recognised:

- User terminal mobility—Each user terminal will continue to use its own IP address when connected to the Internet through the GMMT. Mobility must then be managed between the GMMT and the user's home network.

- GMMT mobility—Whenever the mobile IP-based GMMT changes its point of attachment to the Internet, for example, because of an ISHO, an "update" of the CoA associated to this GMMT must be performed. The HA must be informed to forward the packets to the new location.

In order to develop a protocol stack capable of providing QoS and mobility to the GMBS, each level has to be appropriately managed by the GMMT and by the other network elements of the GMBS network.

To better understand the needed interactions, and to develop a mobility management scheme able to exploit the features of the GMBS, both the GMMT structure and the F-ISP network structure need to be examined.

A possible structure of the GMMT is shown in Figure 6.2. Each terminal has an independent access to the radio subsystem. Location information is available to the T-IWU that implements the multi-segment interface depicted in the same figure. The IP layer of the GMMT interfaces with each segment and with the Global Positioning System (GPS) module by means of the T-IWU.

Another important point to examine is the structure of the network that provides the radio coverage: Different segments interact in order to deliver to end-users the expected QoS. These interactions need to be supported by the transport network that connects the different segments in a "F-ISP Network". The control information that is added to the data flow has to be as small as possible in order to reduce the bandwidth required to uphold each connection.

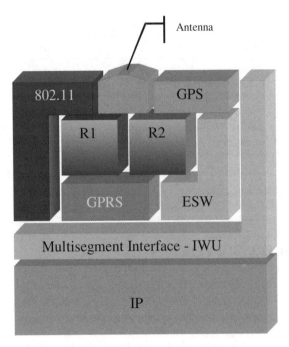

Figure 6.2   A possible GMMT structure ($R1$ and $R2$ represent the Uplink/Downlink radio front-ends).

Two possible approaches can be considered for the structure of the network, the first one solves mobility issues related to the GMMT inside the F-ISP network while the second one relies also on external objects for performing this task.

If the first approach is chosen, it implies that mobility issues are tackled inside the wireless access network and that QoS and mobility must be transparently managed. This is required to allow the rest of the network to be independent of the structure of the access network and of the actions needed to support mobility. A structure that corresponds to this approach is shown in Figure 6.3. The Bandwidth Brokers (BB) manage the resources inside each sub-network. There are different levels of HAs each one manages the mobility for a different entity, depending on the networks they belong to.

On the other hand, one can choose to depend upon mobility management entities located outside the F-ISP network. In this case, a network architecture that is able to support mobility using a distributed architecture must be used. Following this approach, a part of the objects involved in the protocols would not be under the control of the F-ISP network and it will be impossible to require them to respect particular constraints. For example, if an external node is involved in the management of an IP-level handover, it may not be possible to control the time needed to complete this procedure.

To reduce control information and to optimise the use of the network infrastructure, it may also be important to exploit the location information provided to the GMMT by its GPS receiver. This information is useful both to determine the position as well as the level of mobility of the GMMT, and to create coverage maps that could be used to improve or optimise network infrastructures.

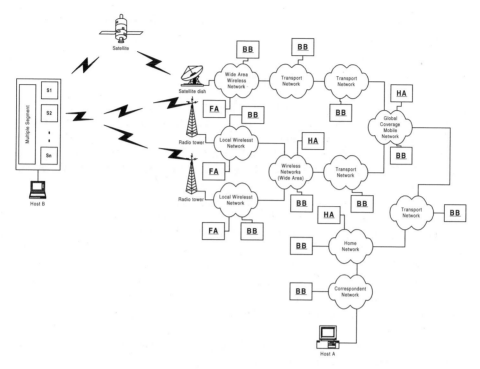

Figure 6.3 A possible network structure.

6.4.2 Mobile Nodes and Mobile Router Issues

The GMBS envisages different kinds of GMMTs, some of which are personal devices, while others are designed to be carried by a vehicle (train, bus, aircraft or even a car) and can be connected using a local area network (LAN) to the laptops used by the passengers. A distinction has to be taken into account depending on the terminal type.

For a personal device, the terminal can be considered a mobile host of an inter-network. In this case, the terminal has to support and provide service to *a single user*. This implies certain bandwidth requirements and a protocol stack being able to manage one single user.

In the second case, collective vehicular, the terminal is a mobile router. The terminal has to support *many users*, and must also manage the available bandwidth in order to fairly share it between the users. It is also important to be able to support the terminal mobility because it is likely that a host connected to a mobile router will be a laptop (for example on a train it could be a computer belonging to a passenger). In this case, the GMMT should support Mobile IP as a Foreign Agent (FA) (for Mobile IPv4) in order to allow the laptops of the users (connected by means of a LAN) to use their personal IP addresses. Also DHCP should be supported for those users that are not interested in using a personal IP address but only in connecting to the Internet with their own terminal.

The mobility management of the "multi-user" terminal as seen by the GMBS is handled with the same procedures as those relevant to the "single user". The difference between these two kinds of terminal is that since the "multi-user" terminal has to support more than one user, it has to handle more than one session in parallel, in addition to the management of the LAN interface that is used to connect the terminals belonging to the users and to the GMBS multi-user terminal.

6.4.3   Mobile IP Over a Multi-Segment Network

In order to provide both mobile and seamless connectivity to the Internet for the GMMT, it is important to solve two problems: The first one is to transmit (and receive) the information using a mobile radio link that is limited both in coverage area and transmitted power level. The second is to take the IP packets addressed to the mobile nodes (laptops connected via the LAN to the GMMT) up to a wireless link able to reach the correct GMMT. Mobile IP is used to address the second issue.

Mobile IP is a set of protocols designed to provide a certain level of mobility to Internet terminals. Using Mobile IP it is possible to disconnect a laptop from one office LAN and reconnect it to another LAN and still use the same IP address (Home IP Address—Personal IP Address).

If the above LANs are wireless LANs (W-LAN) and provided a handover procedure is used to integrate Mobile IP with the available link layers, then a laptop will be able to move from an office to another by changing its CoA along the way but maintaining its home address and allowing the applications to continue to work. This type of handover is referred to as the IP-level handover.

The IP-level handover is managed directly by the user-terminal (the mobile node), making it possible to reduce the amount of work needed at the network side to decide whether or not a link within the access network has to be discarded and replaced by another one.

In a multi-segment scenario, integration of satellite links (M-ESW), cellular networks (GPRS, UMTS) and W-LANs is required. For M-ESW and GPRS/UMTS, wide-area coverage is provided by means of low level handover functions whereby handover is managed transparently for the upper layers.

6.4.4   Handover Issues

Intra-Segment Handover

Handover occurs when the GMMT begins to leave the area covered by the current segment and moves into a second area, where coverage is provided by another network of the same segment. Although the radio segment is the same, a different network provides the connection. In this case, intra-segment IP-level handover takes place in order to route IP packets addressed to that GMMT (Mobile Node) to a point of the Internet where they can be routed to the correct link.

A different kind of intra-segment handover is that performed by a cellular network when the GMMT moves from one cell to another. In this case, the handover is transparent to the IP layer which in theory could continue to use the same connection.

It is important to distinguish between the low level intra-segment handover, which utilises segment specific protocols, and an IP level intra-segment handover. IP-level intra-segment handover enables MTs to receive/transmit packets through the foreign network which is closer to the terminal's actual location, hence reducing delay and network load. High level intra-segment handover will also make it possible to support the handover between different networks belonging to the same segment that do not have another way to perform the handover, such as the case in W-LAN.

For GPRS/UMTS and M-ESW, intra-segment handover will follow segment specific procedures, which are completely managed by the lower layers of each segment. Hence they will not be discussed further. However, for W-LAN, only IP-level intra-segment handover is considered. During or after the intra-segment handover, the GMMT will have to perform two functions:

1. Updating of the CoA assigned to the GMMT every time it gets connected via a certain segment: This is needed in order to let the GMMT continue to receive the IP packets addressed to the mobile nodes when connected to the new network.

2. Resource request on the new link to ensure IP-QoS (for each QoS session) is granted on the new route between mobile nodes (connected to the GMMT) and each correspondent node.

Inter-Segment Handover

Although intra-segment handover could be managed at link layer in certain networks, it may not be feasible to provide ISHO at link-layer function in the multi-segment network being considered. It requires a centralised object having the knowledge that a GMMT can be reached at link layer level through its specific segment. However, it is rather complicated to develop such a centralised object because it would require real-time access to the internal databases of each radio segment. Nevertheless, it may be possible to develop at least a statistical knowledge of the coverage in each segment and use this knowledge to assist the handover using proper IP-based signalling for the GMMT.

In order to perform this, a specific functional layer has to be developed on top of the protocol stack of each access segment in order to support the ISHO at the IP level. In fact, managing ISHO at the IP level is the most important task assigned to the T-IWU.

According to the Internet paradigm, the T-IWU could be considered as a router because it is a host connected to more than one link layer. This implies that the T-IWU has to perform IP router functions to connect different links together without making particular assumptions on the secondary functions of each link. Nevertheless, there is also a big difference between the T-IWU and an ordinary router. The T-IWU will decide on the most suitable link to send a packet according to radio link quality and/or to QoS considerations, while a router will make this decision according to the link cost functions on the path between the originator and the destination of the packets by means of a routing algorithm.

Hence, the differences lie in the software component that rules the decisions made by the T-IWU to enable it to support the ISHO through an interaction with the hardware of each terminal segment.

6.5 GMBS INTER-SEGMENT MOBILITY PROCEDURES

6.5.1 Overview

The ISHO procedures discussed in this section assume the following: The GPRS segment of the GMBS refers to the second-generation GPRS (2G-GPRS) without the corresponding 2G circuit-switched system GSM. In the scope of the GMBS, release 98 of GPRS is considered [ETS-02a]. The term handover is not used in 2G-GPRS, since the relevant terms for mobility with (and without) an active session are "cell update" and "routing area update". In the

GMBS terminology these procedures are of the type intra-segment handover. The UMTS segment of the GMBS refers to the third-generation GPRS (3G-GPRS) without the corresponding 3G circuit-switched domain. Thus the reader should be aware of the differences in terminology:

- GPRS in the standards world refers to packet services for both 2G and 3G, here it refers only to the 2G.

- UMTS in the standards world refers to both circuit-switched and packet-switched domains, here it refers to the packet-switched part only.

Problems with ISHO mainly arise due to different link quality measurements and the different frequencies used by the two systems. Besides this, problems also arise due to the different delays on the satellite and terrestrial segment. Therefore, co-operation between the terrestrial and satellite access segments is of vital importance to ensure the smooth operation of the integrated network. There are two possible ISHOs:

- Handover from terrestrial to satellite system.

There are two reasons for triggering this type of handover: (a) when the GMMT is going outside the terrestrial coverage; (b) when the GMMT, registered within the terrestrial segment, is trying to enter a blocked cell (completely busy cell). In both cases, satellite coverage is obviously needed. Therefore, in order to perform this task, the terminal must be made aware, primarily, of the satellite availability in terms of signal strength and satellite radio link availability and then, if possible, it should also be informed about satellite resource availability. The GMMT can perform the signal measurements and check the availability of physical channels.

- Handover from satellite to terrestrial system.

This type of handover will be triggered when: (a) the GMMT re-enters the terrestrial coverage; or (b) the GMMT is leaving a blocked cell, as this is the reason for using the satellite resources even if the GMMT is within the terrestrial system coverage. In order to perform this task, the GMMT must be made aware, primarily, of the terrestrial system availability in terms of signal strength and terrestrial radio link availability and then, if possible, it should also be informed about terrestrial segment resource availability. The GMMT can perform the signal measurements and check the availability of a physical channel on the selected cell.

Other types of information, such as the position of the terminal, could be considered. If the information is accurate, the terminal would be able to predict or determine the availability of the terrestrial segment without measuring the availability of the radio link. This information could then directly be used to make the handover decision. In addition, the handover algorithm can also take into account the QoS perceived by the user, the battery status and other relevant network parameters.

6.5.2 Handover Strategies

When considering handover strategies, it is important to bear in mind the three phases during handover, namely, the handover initiation, decision and execution phases. The initiation

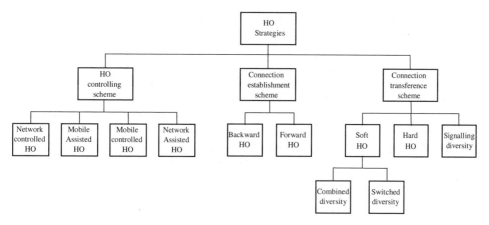

Figure 6.4 Overview of the handover strategies.

phase involves monitoring functions, data collection and processing. The decision phase involves the subsequent decision making process on whether to initiate the handover upon reception of the report from the initiation phase. Finally, the execution phase involves path creation and switching, handover completion and route optimisation.

Handover strategies refer to the schemes adopted in each of the three handover phases. Figure 6.4 describes an overview of the handover strategies.

In the above figure, the handover controlling schemes and the connection establishment schemes will be applied during the handover initiation and handover decision phases, while the connection transference schemes are used in conjunction with the handover execution phase.

There are four handover controlling schemes: mobile controlled handover (MCHO), mobile assisted handover (MAHO), network controlled handover (NCHO) and network assisted handover (NAHO). Their characteristics are summarised in Table 6.2. A more detailed description of each of these schemes can be referred to in [SHE-01].

Handover connection establishment schemes are responsible for setting up the new signalling channel and the disconnection of the old signalling links. There are two ways in which this can be performed, either adopting the backward handover or the forward handover scheme. In backward handover, the old signalling link is used for the exchange of handover signalling messages until the handover procedure is completed. Backward handover may not be reliable when the propagation environment suffers from serious shadowing. In these circumstances, the handover process will have to be terminated as signalling exchange will no longer be possible. In order to increase the reliability, the forward handover scheme adopts a different approach whereby a new signalling link is established and used for

Table 6.2 Characteristics of handover controlling schemes

	MCHO	NCHO	MAHO	NAHO
Handover process	Centralised	Centralised	Decentralised	Decentralised
Terminal complexity	High	Low	Moderate	High
Speed of handover	Fast	Slow	Slow	Fast
Signalling load	Low	Low	High	High
Reliability	Moderate	Moderate	High	High

the exchange of handover signalling messages. Information on the old connection is used to transfer the connection to the new one. Once everything is firmly established, the old network resources are released.

There are basically two classes of handover when referring to the connection transference scheme: soft handover and hard handover. These schemes are related to the setting up of new user traffic channels and the release of the old ones. Soft handover is also known as diversity handover. Another variant of soft handover named signalling diversity is also included in the figure.

In soft handover, the current link and the target link are both maintained before the execution of the handover. This is an example of a seamless handover where the service provided during the handover execution appears to be unaffected by the handover process. Soft handover can be further divided into two classes, switched diversity and combined diversity. Switched diversity refers to the handover process where communications are established through one of the open links, but not by both simultaneously, whereas combined diversity occurs when both links are used for communication during the handover.

Hard handover occurs when the old connection is broken or released before the new connection is activated. This usually results in a short interruption of service. Systems implementing a hard handover algorithm usually operate in non-seamless mode. GPRS adopts hard handover operation. Hard handover is applicable for ISHO, but it is not very practical for satellite systems due to the long propagation delay.

The implementation of soft handover for ISHO is rather complicated and requires careful consideration, in particular, where the software algorithm and synchronisation is concerned. Due to the propagation delays on the original connection and the target connection, the implementation of soft handover for ISHO is foreseen to be very difficult.

However, it is also impractical to implement a hard handover scheme, as handovers are required to be as seamless as possible. Hard handover results in a short interruption of service resulting in the degradation of performance, which is not desirable in future mobile communications systems. In [EFT-98], a different type of handover connection transference scheme is adopted, known as signalling diversity.

Recall that in soft handover, both the current link and the target link are kept open during the handover process. This method is not very suitable for ISHO due to synchronisation problems and the large delay difference between the two environments. Signalling diversity is a variant of the soft handover. Under this scheme, a new signalling link is established between the GMMT and the target network for the establishment of a new traffic channel while both the old signalling and traffic links are still in use. Once the new traffic link between the GMMT and the target network is firmly established, the old signalling and traffic links are dropped. Therefore, no synchronisation is required for this scheme as both the old and new connections serve different purposes. Using this scheme, the handover break duration is reduced to the propagation delay difference between the satellite and terrestrial networks.

6.5.3   GMBS Inter-Segment Handover Approach

ISHO in GMBS can be foreseen as being controlled by a common control unit referred to as the "Terminal Inter-Working Unit" (T-IWU), as shown in Figure 6.5. As such, the following approach is proposed for the GMBS ISHO.

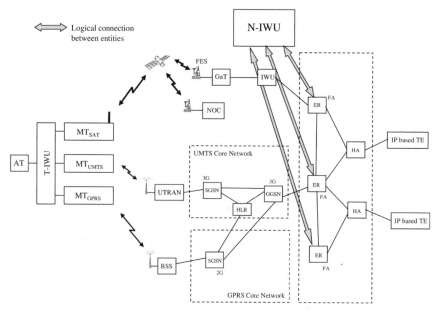

Figure 6.5 GMBS integration scenario.

- *Handover Controlling Scheme*: Initially, both the MCHO and MAHO are considered to introduce some flexibility in the design. However, MCHO is the primary controlling scheme to ensure that the modification done on the network is as minimum as possible. On the other hand, the feasibility of employing MAHO in the GMBS is also considered as well in parallel with the implementation using MCHO. MAHO is chosen to avoid too much complexity in the GMMT and to increase the performance of the system in terms of handover decision accuracy.

- *Connection Establishment Scheme*: Forward Handover is chosen. This connection establishment scheme is considered most suitable for ISHO from the satellite system to the terrestrial system. The probability of a sudden drop in the satellite link is more likely to occur when backward handover is employed, due to the on-off behaviour of the satellite link. However, backward handover might be considered for the ISHO from the terrestrial system to the satellite system.

- *Connection Transference Scheme*: The signalling diversity scheme is selected because it allows a reduced break within the call when performing the handover. It also has the advantage that synchronisation is not needed between the link to the terrestrial system and the link to the satellite system (which is difficult due the significant difference in the propagation delay of the two systems), when performing an ISHO between the terrestrial segment and satellite segment.

The design of the GMBS ISHO algorithm is based on the above selected schemes.

As the MCHO scheme is to be employed, each segment specific MT makes the downlink measurements periodically. These measurements are sent to the functional entities related to inter-segment mobility in the T-IWU (see Chapter 7).

Handover decisions are made by the T-IWU. If no intelligence is employed in the network, the handover decision is based on the radio link parameters and user profile parameters. However, if some intelligence is incorporated in the network, other parameters such as the network conditions can be taken into account.

The MT selects the target segment based on the criteria and measurements mentioned above. Then it triggers a GMBS ISHO procedure.

The MT requests for the needed bearer resources within the target segment. However, this signalling is done within the new segment as forward handover is employed.

When the new resources are available, handover is executed. Here, the MT will obtain a new CoA from the new segment and then registers this CoA with its HA and correspondent Node. This is also known as Mobile IP Registration/Notification.

The functional entities related to inter-segment mobility located in the T-IWU inform the old segment that the resources associated to the old connection can be released. This is to overcome the problem related to the release of resources when the old link is lost before completing the handover procedure.

6.5.4   Fuzzy Logic Based Handover Algorithm

6.5.4.1   Handover Algorithm Design Consideration

One of the main conditions in designing a handover and segment selection algorithm is the determination of the handover criteria used in the algorithms. This is necessary as the criteria are responsible for determining the most suitable segment for a user and the necessity for initiating a handover. This section aims to describe a new segment selection and handover algorithm that can be implemented for GMBS. Essentially, the criteria for both algorithms are similar. However, since the emphasis for mobility management is more inclined towards the handover algorithm, the discussion on the criteria will be focussed on how the criteria affect the handover algorithm.

Traditional handover algorithms fundamentally depend on the signal strength, power control and traffic of the network, as measured by the MT. This is employed in GSM, where three different types of handover can be performed: rescue handover, confinement handover and traffic handover [MOU-92].

Rescue handover occurs when the terminal is moving out of the radio coverage area of a particular cell, which results in the reduction of the signal strength. Handover is then initiated when the signal falls below the threshold required by the system. When handover is based on power control, handover can be initiated even though the transmission quality of the serving cell is still adequate. The purpose of confinement handover is, therefore, to improve the interference level and also to minimise the transmission power. Finally, traffic handover occurs when congestion occurs in the serving cell.

The handover parameters described above are not sufficient when the user is capable of using real-time applications with different wireless access segments. This is because the criteria only take into account the conditions of the network without considering requirements of the applications utilised by the user. Therefore, no QoS guarantees can be offered for a specific application. In order to counteract this problem, an adaptive priority-based handover algorithm can be implemented, where users are generally given the option to influence the result of handover and segment selection. This method would enable the possibility of initiating a handover or segment selection using 'user defined QoS characteristics' such as perceived QoS, cost, security and priority of the user. In addition to this, 'technology based QoS characteristics' can also be implemented, where the criteria are based on the latency and reliability of the network, and available bandwidth.

Since different objectives and criteria are employed in the system, the proposed solution for both segment selection and handover utilises the fuzzy logic concept, where a robust mathematical framework for dealing with imprecision and non-statistical uncertainty is introduced [GHO-98].

This is advantageous in the GMBS, as a fuzzy logic system is flexible and is capable of operating with imprecise data and can therefore be used to model non-linear functions with arbitrary complexity. Furthermore, since most algorithms related to fuzzy logic use a relativity approach, it reduces some problems faced by designers when comparing dissimilar systems such as the different delays introduced by the systems, different link quality measurements and the different operating frequencies. This concept has been used success- fully in automatic control applications and could be extended to include applications in telecommunications networks.

6.5.4.2 Fuzzy Logic Concept

In fuzzy logic, an event or situation does not have to be either 'true' or 'false'. Instead, these events can be classed as 'quite true', fairly true', 'very true' or 'not true'. In a fuzzy set, an element is related to a set by a membership function μ, which usually takes on a value of between 0 and 1 to represent the degree of membership in a set [EDW-94]. When expressed mathematically, the functional mapping of a fuzzy set is given by

$$\mu_A(x) \in [0, 1] \tag{6.1}$$

where the symbol $\mu_A(x)$ is the degree of membership of element x in a fuzzy set A.

This is different from a classical or crisp set, where an element will only be considered as a member of a class if it has full membership in a set. If an element does not satisfy the conditions of a set, the element will not be included at all. Hence, when a crisp set is used, the membership value of an element in a set can only be assigned a value of either zero or one (false or true). When expressed mathematically,

$$\chi_A(x) = \left\{ \begin{array}{ll} 1, & x \in A \\ 0, & x \notin A \end{array} \right\} \tag{6.2}$$

where the symbol $\chi_A(x)$ gives the indication of an unambiguous membership of x in set A.

As an example, if signal strength is considered in a crisp set, the signal can only be considered to be either "strong" or "weak" and not both simultaneously, whereas in a fuzzy set, the signal can be classed as "quite weak" or "not so strong" or "medium". This indicates that an element in a fuzzy set can have membership in more that one set. The membership values are obtained by mapping the values obtained for a particular parameter onto a membership function. The most commonly used membership functions are in the unit interval 0 to 1. Several types of membership functions can be implemented in a system, as shown in Figure 6.6.

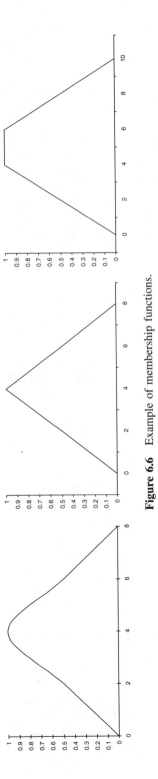

Figure 6.6 Example of membership functions.

Several fuzzy set operations are important when considering the implementation of fuzzy logic in GMBS. The two most common operations are intersection and union operation. The AND (intersection) of any two fuzzy variables results in the minimum truth value, whereas the OR (union) of two fuzzy variables results in the maximum truth value.

$$\mu_{A \cap B} = \min(\mu_A, \mu_B)$$
$$\mu_{A \cup B} = \max(\mu_A, \mu_B)$$

(6.3)

Before applying the fuzzy logic into the design of the GMBS handover algorithm, two issues need to be considered:

- Handover requirements;

- Segment selection and handover criteria.

6.5.4.3   Handover requirements

In UMTS, several types of handover are possible as outlined in [3GPP-00]. In the specifications, handover is defined as the ability to maintain continuity of service when moving between different radio coverage areas. Handover also occurs when the radio resources providing a service change from one UTRA (UMTS Terrestrial Radio Access) mode to another (UTRA-FDD to UTRA-TDD and *vice-versa*). In addition, [3GPP-00] also highlights several important parameters for the requirements of handover between different radio systems, in particular on the handover between UMTS and UMTS, UMTS to GSM and GSM to UMTS. The general principles governing handover identified in the document are as follows:

- The continuation of the active calls;

- Charging, billing and accounting for handover;

- The ability to check whether handover can be granted by the home network and accepted by the target network;

- Initiation of handover only if the target network can provide the radio channel required by the terminal;

- Avoidance of successive handovers between networks for the same call;

- The possibility of notifying the user of the handover (for charging purposes);

- The handover shall not compromise the security of the network providing the new radio resources;

- Degradation of service during handover (from UMTS to GSM) should not be worse than during intra GSM handover;

- Duration of discontinuity for packet-switched and circuit-switched should be shorter than that for GSM circuit-switched speech calls;

- Indication of minimum acceptable QoS for services after handover.

These requirements will be used as the baseline for designing the ISHO algorithm.

6.5.4.4   Handover Criteria

In the following, the different parameters which may trigger a handover are discussed. These parameters fall into three main categories:

- Current Network Conditions
- User-based QoS characteristics
- Technology-based QoS characteristics

Current Network Conditions

Signal Strength

Signal strength has been the main criterion in initiating handovers in many traditional circuit-switched systems, such as GSM, where the downlink and uplink transmission quality in terms of raw bit error rate (BER) is measured by the MS and base transceiver station (BTS), respectively. Even though other parameters have been introduced for the design of the algorithms, this parameter still remains one of the most influential. In order to enable the design and simulation of the segment selection and handover algorithms, it is important that the correct model for the radio propagation in all the different segments is used. However, for simplicity, all segment selection and handover criteria will be rated on a scale of 0 to 1, with 0 being the most undesirable condition and 1 being the most desirable. Later this can be changed to reflect the actual network conditions.

Current Network Load on All Reachable Networks

In addition to measuring the signal strength, it is also important to know the current network loads of all reachable networks. In GMBS, this is performed by the QoS Support Module (QASM) (see Chapter 5), which is used to test the condition of all the reachable networks. From measurements provided by the QASM layer, the current network load can be anticipated. Currently, only two measurements are available, packet delay and queue delay. Using this information, together with the size of the packet generated by the QASM layer, a simple algorithm has been designed to determine the condition of all reachable networks.

Current Battery Status and Power Consumption

This parameter is usually not very important when used in 2G systems as only one terminal is used at any one particular time. However, in 3G and fourth-generation systems, where the concept of the multi-mode terminal is used, this criterion could be important in segment selection and handover. Two different methods of representing the level of the battery could be used.

1. as a percentage of the overall capacity of the battery,

2. the amount of time left before the battery runs out.

If the capacity of the battery and power consumption of the terminal is known, the terminal can predict the correct time to perform a handover or segment selection. In order to test the performance of the handover algorithm, it is important that some information on the battery used by the terminal is available.

Technology-Based QoS Characteristics

Network Latency

The network latency can be measured by calculating the transmission time, the response time and the delay jitter. The response time refers to the round trip delay experienced by the packet from the time it was sent until the time a reply is received. The network latency can affect the performance of the system, depending on the type of application used.

Available Bandwidth

The available bandwidth is the amount of data that can be transmitted in a fixed amount of time and is usually measured in bits per second or bytes per second. Bandwidth can be further separated into three different parameters, namely, system level data rate, application level data rate and transaction rate [CHA-99].

The system level data rate refers to the amount of bandwidth that is available or requested by the system. Application level bandwidth is important as it defines the QoS provided to the users. Therefore, if a segment is deemed unsuitable to support a particular type of application, such as video conferencing, the segment selection and handover mechanism will be activated to change to a new segment. Currently, various bandwidth driven applications exist, starting from simple applications such as e-mail and facsimile to bandwidth hungry applications like voice over IP (VoIP) and video conferencing (see Chapter 4 for further information). Table 6.3 shows a comparison of data rates applicable to a GMMT. The third parameter is the transaction rate, which denotes the number of operations requested or processed per second.

From the table, it can be seen that the UMTS segment can offer the highest bandwidth when the segment is available. Therefore, this segment will be the preferred choice if bandwidth is the main priority, followed by the GPRS segment and the satellite segment. However, bandwidth alone is not enough to guarantee the correct choice of segment.

Reliability of Network and Coverage

When the reliability of the network is considered, several parameters can be taken into consideration. In [WAN-99], it was suggested that the reliability can be characterised by the number of retransmissions required by the network. However, reliability can also be calculated by using the mean time to failure, mean time to repair, mean time between failures and availability of the system. The latter parameter can only be used if sufficient data is available for all the access segments.

Network coverage can be another criteria used for segment selection or handover. This is particularly useful if the current location of the terminal is also known. Therefore, a segment

Table 6.3 GMBS user terminal data rate comparison

		Mobile			Portable	Service area
		<10 km/h	<120 km/h	<250 km/h		
Satellite	Individual	Uplink: 160 kbps Downlink: 6 Mbps	Uplink: 160 kbps Downlink: 6 Mbps	N/A	Uplink: 160 kbps Downlink: 6 Mbps	Rural, Suburban/ urban, Short range outdoor/ indoor (via W-LAN), Polar zones (via LEO satellites)
	Collective	Uplink: 0.512/2 Mbps Downlink: 16 Mbps	Uplink: 0.512/2 Mbps Downlink: 16 Mbps	Uplink: 2 Mbps Downlink: 16 Mbps	Uplink: 0.512/2 Mbps Downlink: 16 Mbps	
GPRS	Individual	171.2 kbps	171.2 kbps	171.2 kbps	171.2 kbps	Suburban/ urban,
	Collective[a]	171.2 kbps	171.2 kbps	171.2 kbps	171.2 kbps	Short range outdoor/indoor
UMTS	Individual	2 Mbps	384/512 kbps	144/384 kbps	2 Mbps	Suburban/ Urban,
	Collective[b]	2 Mbps	384/512 kbps[a]	144/384 kbps	2 Mbps	Short range outdoor/indoor
W-LAN[c]	Individual	11/5.5/2/1 Mbps	11/5.5/2/1 Mbps	11/5.5/2/1 Mbps	11/5.5/2/1 Mbps[d]	Short range outdoor/indoor
	Collective	11/5.5/2/1 Mbps	11/5.5/2/1 Mbps	11/5.5/2/1 Mbps	11/5.5/2/1 Mbps[d]	Short range outdoor/indoor

[a]Higher bit rates may be obtained by connecting a number N of GPRS MTs to the T-IWU.
[b]Higher bit rates may be obtained by connecting a number N of UMTS MTs to the T-IWU.
[c]The W-LAN is used to bridge the satellite connectivity in shadowed areas.
[d]No speed limitation exists if the terminal remains in the same cell; stringent requirements for maximum speeds across cells' boundaries have to be met by the W-LAN product.

selection and handover can be initiated when the terminal detects that the user is moving out of the network coverage area. However, to use this parameter, sufficient data must also be available for the access segments.

User-Based QoS Characteristics

User Preferences/Priority

This criterion allows the user to decide whether any of the segments has a higher priority than the rest. In GMBS, the user is given a choice of priority segments and prohibited segments. These segments are then given weightings based on the choice of the user.

QoS perceived by the User

This criterion gives an indication of the QoS observed by the user. The users could be given several options in this criterion, such as the detail of the picture received, colour accuracy, video rate, video smoothness, quality of audio, throughput and video/audio synchronisation [CHA-99]. These parameters can be expressed in terms of pixel resolution, jitter rate, audio sampling rate, number of bits and video and audio stream synchronisation. Therefore, handover can be initiated when the user detects any degradation in any of the parameters mentioned above. This functionality is implemented in the entity called Quality Evaluation Function (see Chapter 7 for a description of this entity).

Service Provision

Service provision is important in a multi-segment environment as it matches the most suitable segment with the service requested by the user. When segment selection and handover are triggered due to service provision, this indicates that the segment currently in use is not able to provide the required service or QoS required by the user. For example, if the current segment is GPRS and a video streaming application is required, handover or segment selection might be initiated to change the current segment from GPRS to another segment that is capable of supporting video streaming. Therefore, the types of service required by the user have a great influence on the selected segment.

Mobility of User

In order to ensure that the algorithm is as accurate as possible, it is also useful to know the mobility of the user. Therefore, if the user has high mobility, it is more advantageous to choose a segment with a wider coverage. This reduces the frequency of handover and thus makes the algorithm more efficient. This is also important if the network coverage can be used as one of the parameters used for handover. By considering the mobility of the user and the network coverage map, the call dropping probability can be estimated and handovers can be initiated before any calls are dropped.

Charging Models

In modern telecommunication systems, the charging model can be very critical for any handover or segment selection algorithm. Two different charging models can be employed; the per-use cost or the per-unit cost. When per-use cost is used, a user is only charged when establishing connection or when gaining access, whereas per-unit cost refers to models that charges according to the amount of data sent or the amount of time online. In the GMBS, it is foreseen that the per-unit cost will be employed. For the satellite segment, the user will be charged for the amount of time the user is online, whereas for GPRS and UMTS, the charging model is likely to be volume-based, where users are charged based on the volume of data/packets sent and received. This is described in [3GPP-99] for UMTS. Therefore, when designing the algorithm, it is assumed that the GPRS has the lowest cost, followed by the UMTS and the satellite segment.

History of User's Activities

Another criterion that can be used is based on the history of a user's activities such as the duration, bytes transferred and observed throughput. This involves keeping a record of all user activities and then deciding on the best segment for the user. However, this method would increase the complexity of the handover algorithm tremendously, as the algorithm would need to incorporate a significant amount of signal processing.

6.5.4.5   Handover Algorithm

Overview

The handover algorithm presented in this section will encompass the handover initiation and decision phases. Handover initiation is one of the most important components of an ISHO since it influences the success or failure of a handover procedure. The decision to initiate a handover depends on a number of control elements; in particular, on the way a signal is averaged over time and then compared between two or more elements. This is described in the following section. In terms of QoS, two approaches are considered during handover initiation. In the first approach, handover is initiated when the QoS perceived by the user degrades and falls below a specified threshold. Degradation is detected based on the speed of transmission and the packet loss ratio perceived by the user. The second approach is related to service provision. Here, handover is initiated when the user requests for additional services, which could not be provided by the current segment.

The criteria used for handover initiation are as follows:

- Signal Strength

- Reliability Model

- Network Coverage

- QoS Perceived by the User

In the decision phase, the system is required to select the most suitable segment based on the information provided by both the system and the user. The algorithm will be based on fuzzy logic. Figure 6.7 shows the block diagram for the GMBS ISHO algorithm. Its detailed description will be given in later sections.

An important point to note when considering handover is the frequency of handover which is in turn governed by the signal averaging interval. Too short an averaging interval may

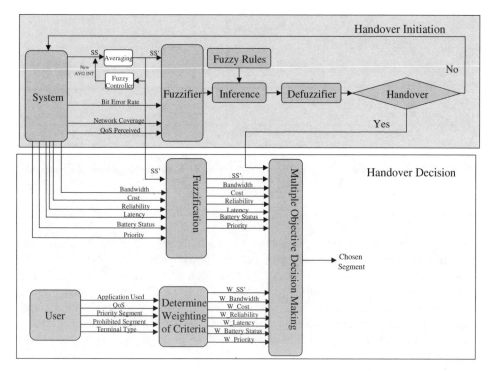

Figure 6.7   GMBS Inter-segment handover algorithm. © IEEE 2001.

result in unnecessary signalling load. Conversely, too long an interval may result in dropped calls.

Hence, before presenting the handover algorithm, a methodology, which is also based on fuzzy logic, used to determine the signal averaging interval is discussed.

Application of Fuzzy Logic for Signal Averaging Interval

Averaging a signal is usually necessary to remove rapid fluctuations due to multipath fading [POL-96]. This is to ensure that the handover rate is at an acceptable level and prevents a user from being transferred back and forth from one segment to another. However, a question then arises as to the size and length of the averaging window. In [MUR-91], it was suggested that for a system with a relatively smooth path loss characteristic, the averaging interval should be large enough to remove variations due to fading, whereas for micro-cellular systems, a long averaging interval is not desirable due to sudden path loss as a result of the corner effect. The approach adopted for GMBS involves averaging the signal received with different window size and length for each segment using fuzzy logic principles for both satellite and terrestrial components.

The algorithm for determining the averaging interval using fuzzy logic was introduced in [LAU-95] and is known as the Fuzzy Adaptive Averaging Interval and Hysterisis. In this algorithm, the averaging interval is varied based on the relative signal strength and

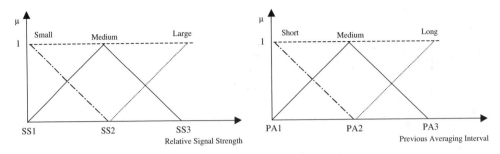

Figure 6.8 Membership functions of relative signal strength and previous averaging interval.

the previous averaging interval. Two membership functions are used to represent the relative signal strength and the previous averaging interval, as shown in Figure 6.8.

The system designer can set the values SS1, SS2, SS3, PA1, PA2 and PA3. This is useful when comparing signals from different segments as the measurement of signal strength might vary depending on the types of system employed. Using this approach, the relative signal strength can be assigned membership values in the corresponding classes (i.e. small, medium or large for relative signal strength). The same principle applies for the previous averaging interval.

Once the classification of the relative signal strength and averaging intervals have been defined, nine fuzzy control rules can be applied to these two parameters, as shown in Table 6.4. These rules are determined by the system designer and can be changed to suit the requirements of each segment.

This algorithm is logical as the averaging interval is longer when the relative signal strength is small. This is designed to counteract the 'ping pong' effect experienced by the two segments when the signal levels between both segments are nearly equal.

However, this algorithm is only useful when two segments are considered. When more segments are used, the algorithm will have to be modified to take into account the increased number of segments. In cases like this, it is possible to perform a pairwise comparison of the

Table 6.4 Fuzzy rules for determining averaging interval

IF Relative signal strength	AND Previous averaging interval Is	THEN Averaging Is
LARGE	LONG	DECREASE
LARGE	MEDIUM	DECREASE
LARGE	SHORT	DECREASE
MEDIUM	LONG	CONSTANT
MEDIUM	MEDIUM	CONSTANT
MEDIUM	SHORT	CONSTANT
SMALL	LONG	INCREASE
SMALL	MEDIUM	INCREASE
SMALL	SHORT	INCREASE

current and target signal strength. The values SS1, SS2, SS3, PA1, PA2 and PA3 vary according to the conditions of the different segments.

Handover Initiation

The handover initiation algorithm, as shown in the upper half of Figure 6.7, is based on the algorithm described in [DAN-00, EDW-00].

From the diagram, it can be seen that the handover initiation algorithm is separated into three stages. In the first stage, the inputs from the system are fed into a fuzzifier which transforms all the different criteria into fuzzy sets. Then, the sets are passed through an inference mechanism, which changes the values of the fuzzy set based on a certain predefined fuzzy set. The function of the defuzzifier is to transform the fuzzy set back into normal values.

In order to decide whether a handover is needed, the current characteristics of the active segment are monitored. However, the current conditions have to be fuzzified into fuzzy sets. This can be performed using membership functions. Different membership functions are defined for different criteria. For simplicity, the membership functions shown in the previous section can be utilised.

For example, when signal strength is considered, it can be evaluated with reference to three different membership sets; strong, weak or medium. Before determining the membership values in each set, a database containing the acceptable threshold for weak, medium or strong for that particular segment is loaded into the system. This could be represented graphically in Figure 6.9(a), where the three lines represent the range available for weak,

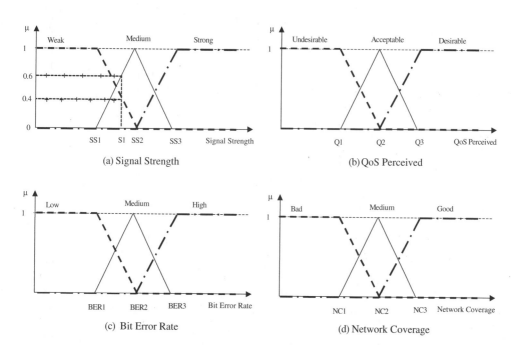

Figure 6.9 Triangular function for handover initiation. © IEEE 2001.

medium and strong. SS1, SS2 and SS3 represent the threshold for weak, medium and strong, respectively. Then, by mapping the position of the received signal strength onto the graph of the membership function, μ, the signal will be allocated with a membership value in each set ranging from 0 to 1. Therefore, if a signal falls between SS1 and SS2, it could be assigned a value of 0 for the strong set, 0.6 in the medium set and 0.4 in the weak set. Figures 6.9(b), 6.9(c) and 6.9(d) refer to the QoS Perceived, Reliability (BER) and Network Coverage, respectively. These membership values will later be used to obtain the handover initiation factor, which determines whether a handover is initiated.

Following this, fuzzy rules, which are formed by using all possible combinations of the different criteria employed for handover initiation, are implemented in the system. These rules are conditional statements that specify how the fuzzy system works and can be implemented in a look-up table. An example of the table is shown below. In the table, the results of the combination is represented by Y(Yes), PY(Probably Yes), PN(Probably Not) and N(No). Note that not all the criteria specified have to be used to make up the IF-THEN rules.

IF signal strength is strong, *AND* QoS is Desirable, *AND* Bit Error Rate is medium, *AND* Network Coverage is medium, *THEN* handover = *NO*.

Table 6.5 demonstrates how the fuzzy rules govern the decision for handover initiation.

Table 6.5 Fuzzy rules for handover initiation

Signal strength	Perceived QoS	Bit error rate	Coverage	Result
Weak	Undesirable	High	Bad	Y
Weak	Undesirable	High	Medium	Y
Weak	Undesirable	High	Good	Y
Weak	Undesirable	Medium	Bad	Y
Weak	Undesirable	Medium	Medium	Y
Weak	Undesirable	Medium	Good	Y
Weak	Undesirable	Low	Bad	Y
Weak	Undesirable	Low	Medium	Y
Weak	Undesirable	Low	Good	Y
Weak	Acceptable	High	Bad	Y
Weak	Acceptable	High	Medium	Y
Weak	Acceptable	High	Good	Y
Weak	Acceptable	Medium	Bad	Y
Weak	Acceptable	Medium	Medium	Y
Weak	Acceptable	Medium	Good	Y
Weak	Acceptable	Low	Bad	Y
Weak	Acceptable	Low	Medium	Y
Weak	Acceptable	Low	Good	Y
Weak	Desirable	High	Bad	PY
Weak	Desirable	High	Medium	PY
Weak	Desirable	High	Good	PY
Weak	Desirable	Medium	Bad	PY
Weak	Desirable	Medium	Medium	PY
Weak	Desirable	Medium	Good	PY
Weak	Desirable	Low	Bad	PY
Weak	Desirable	Low	Medium	PY
Weak	Desirable	Low	Good	PY

Table 6.5 (*Continued*)

Signal strength	Perceived QoS	Bit error rate	Coverage	Result
Medium	Undesirable	High	Bad	PY
Medium	Undesirable	High	Medium	PY
Medium	Undesirable	High	Good	PY
Medium	Undesirable	Medium	Bad	PY
Medium	Undesirable	Medium	Medium	PY
Medium	Undesirable	Medium	Good	PY
Medium	Undesirable	Low	Bad	PY
Medium	Undesirable	Low	Medium	PY
Medium	Undesirable	Low	Good	PY
Medium	Acceptable	High	Bad	PY
Medium	Acceptable	High	Medium	PY
Medium	Acceptable	High	Good	PY
Medium	Acceptable	Medium	Bad	PN
Medium	Acceptable	Medium	Medium	PN
Medium	Acceptable	Medium	Good	PN
Medium	Acceptable	Low	Bad	PN
Medium	Acceptable	Low	Medium	PN
Medium	Acceptable	Low	Good	PN
Medium	Desirable	High	Bad	PY
Medium	Desirable	High	Medium	PY
Medium	Desirable	High	Good	PY
Medium	Desirable	Medium	Bad	PN
Medium	Desirable	Medium	Medium	PN
Medium	Desirable	Medium	Good	PN
Medium	Desirable	Low	Bad	PN
Medium	Desirable	Low	Medium	PN
Medium	Desirable	Low	Good	PN
Strong	Undesirable	High	Bad	PY
Strong	Undesirable	High	Medium	PY
Strong	Undesirable	High	Good	PY
Strong	Undesirable	Medium	Bad	PY
Strong	Undesirable	Medium	Medium	PY
Strong	Undesirable	Medium	Good	PY
Strong	Undesirable	Low	Bad	PY
Strong	Undesirable	Low	Medium	PY
Strong	Undesirable	Low	Good	PY
Strong	Acceptable	High	Bad	PN
Strong	Acceptable	High	Medium	PN
Strong	Acceptable	High	Good	PN
Strong	Acceptable	Medium	Bad	PN
Strong	Acceptable	Medium	Medium	PN
Strong	Acceptable	Medium	Good	PN
Strong	Acceptable	Low	Bad	N
Strong	Acceptable	Low	Medium	N
Strong	Acceptable	Low	Good	N
Strong	Desirable	High	Bad	PN
Strong	Desirable	High	Medium	PN
Strong	Desirable	High	Good	PN

Table 6.5   (*Continued*)

Signal strength	Perceived QoS	Bit error rate	Coverage	Result
Strong	Desirable	Medium	Bad	N
Strong	Desirable	Medium	Medium	N
Strong	Desirable	Medium	Good	N
Strong	Desirable	Low	Bad	N
Strong	Desirable	Low	Medium	N
Strong	Desirable	Low	Good	N

Finally, a handover factor is introduced in which the degree of membership is converted into a handover factor using a weighting matrix. This matrix represents the weighting relevant to Y, PY, PN and N, respectively. The equation used to calculate the handover factor is given by the following equation:

$$\textit{Handover factor} = \frac{\sum D_i \times W_i}{\sum D_i} \tag{6.4}$$

where D_i refers to the degree of membership and W refers to the weighting matrix. If the handover factor exceeds a certain threshold, handover will be initiated. The following illustrates an example of the handover initiation algorithm implementation.

Assume that the GPRS segment is the active segment and has the membership function shown in Figure 6.10. The line with crosses (+) represents the values measured for GPRS in

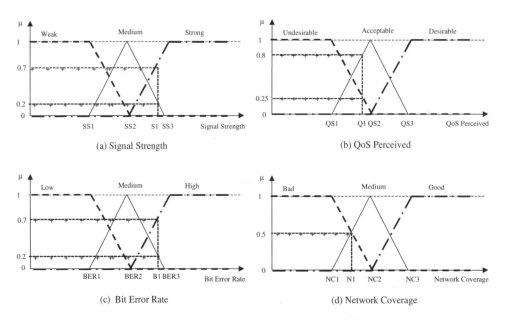

Figure 6.10   Membership function for GPRS.

Table 6.6 Membership values assigned to the measurements obtained from GPRS

Criteria	Weak, undesirable, high or bad (W, U, H, B)	Medium, acceptable (M, A)	Strong, desirable, low, good (S, D, L, G)
Signal strength (SS)	0	0.2	0.7
QoS perceived (QoS)	0.25	0.8	0
Bit error rate (BER)	0	0.2	0.7
Network coverage (NC)	0.5	0.5	0

the current session. As can be seen from the diagram, each measurement can be assigned membership values. These values are shown in Table 6.6.

Once the membership values have been assigned, the sets are compared against the fuzzy logic table defined earlier (Table 6.5) using the IF-ELSE rule. The results are shown in Table 6.7.

Note that since the IF-ELSE rule uses the AND statement, Eq. (6.3) can be used to determine the final membership value in the last column. As can be seen from Table 6.7, the values obtained are actually the minimum values of a particular line.

It can be seen from the table that there are ten decisions with PY and six with PN. When the values assigned to the decisions are observed, it can be seen that for PY, two values are present; 0.2 and 0.25. In such cases, the *UNION* or *OR* function is used, as stated in Eq. (6.3). Therefore, the maximum value (0.25) will be used. The same applies to PN. For PN, 0.5 will be used.

Table 6.7 Result of IF-ELSE rule

Criteria				Decision	Final membership value
SS	QoS	BER	NC		
M(0.2)	U(0.25)	M(0.2)	B(0.5)	PY	0.2
M(0.2)	U(0.25)	M(0.2)	M(0.5)	PY	0.2
M(0.2)	U(0.25)	H(0.7)	B(0.5)	PY	0.2
M(0.2)	U(0.25)	H(0.7)	M(0.5)	PY	0.2
M(0.2)	A(0.8)	M(0.2)	B(0.5)	PN	0.2
M(0.2)	A(0.8)	M(0.2)	M(0.5)	PN	0.2
M(0.2)	A(0.8)	H(0.7)	B(0.5)	PY	0.2
M(0.2)	A(0.8)	H(0.7)	M(0.5)	PY	0.2
S(0.7)	U(0.25)	M(0.2)	B(0.5)	PY	0.2
S(0.7)	U(0.25)	M(0.2)	M(0.5)	PY	0.2
S(0.7)	U(0.25)	H(0.7)	B(0.5)	PY	0.25
S(0.7)	U(0.25)	H(0.7)	M(0.5)	PY	0.25
S(0.7)	A(0.8)	M(0.2)	B(0.5)	PN	0.2
S(0.7)	A(0.8)	M(0.2)	M(0.5)	PN	0.2
S(0.7)	A(0.8)	H(0.7)	B(0.5)	PN	0.5
S(0.7)	A(0.8)	H(0.7)	M(0.5)	PN	0.5

Finally, in order to obtain the handover factor, another weighting matrix has to be constructed, which defines the weighting of each decision matrix. Assume that the weighting is as follows:

$$Y = 0.8, PY = 0.6, PN = 0.4 \text{ and } N = 0.1$$

Then, by using Eq. (6.4), the handover factor is determined.

$$Handover\ Factor = \frac{(0.25 \times 0.6) + (0.5 \times 0.4)}{0.25 + 0.5} = 0.47$$

Since the handover factor is quite small, handover will not be initiated.

Handover Decision

The handover decision phase involves segment selection procedures. In this phase, real-time measurements required to make a handover decision foreseen will be carried out, as there is an active IP connection with the Service Node. Therefore, real-time information can be obtained from this node, enabling a better negotiation of QoS criteria. Figure 6.11, which presents the lower half of Figure 6.7, shows the operations required when using fuzzy logic to perform segment selection during the handover decision.

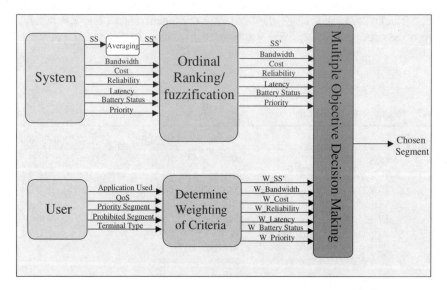

Figure 6.11   Block diagram describing the segment selection algorithm.

It can be seen that the algorithm is separated into two different components, each with a different weighting mechanism. The first component involves the evaluation and comparison of the available segments in terms of each handover criterion and the second component involves evaluating the importance of the criteria, based on the discretion of the network provider and the subscriber.

With the criteria defined above, appropriate objectives can be set for the segment selection algorithm including

- Low Cost;

- Good Signal Strength;

- Optimum Bandwidth;

- Low Network Latency;

- High Reliability;

- Long Battery Life;

- Fulfilment of user defined priority/preferred segment.

From these objectives, the algorithm will endeavour to select the most suitable segment, which can fulfil the seven objectives.

The Ordinal Ranking/Fuzzification block, as shown in Figure 6.11, represents two different approaches in ranking the different criteria in order to select the best segment. The methodology of these two approaches, together with the Multiple Objective Decision Making process will be described shortly.

Crisp Set and Fuzzy Set

It should also be noted that not all the criteria fits into the fuzzy logic concept. Some criteria, such as the segment availability and whether a segment is prohibited have a definite yes or no answer. Therefore, these can be classed under ordinary or crisp sets.

In the GMBS, the operation of the crisp sets might have to be executed first before any other decision making can proceed to determine the most suitable segment. Three of the criteria chosen for segment selection are classed under ordinary sets, whereas seven are classed under the fuzzy sets, as shown in Table 6.8.

Segment availability is considered to be under crisp sets since the availability of the segment can only provide a positive or a negative answer. However, this is an important criterion as the availability influences the ordering of the other parameters. Therefore, if a segment is unavailable, it will be given zero weighting.

Since the user is also allowed to choose a segment that is prohibited, it is also necessary to indicate this parameter in the segment availability matrix. If the user decides to prohibit the

Table 6.8 Ordinary sets and fuzzy sets

Crisp sets	Fuzzy sets
Segment availability	Signal strength
Prohibited segments	Charging model
Terminal types	Available bandwidth
	Network latency
	Reliability
	Priority segments
	Battery status

use of the satellite segment, the weighting of the satellite segment availability will be assigned zero.

The type of terminal used can also be considered to be a crisp set as there is no ambiguity in the type of terminal used. However, this is an important criterion, as it determines the importance of the battery status. Currently, three types of terminal are considered; a car version, large vehicle (collective terminals for public transportation) version and a briefcase version. Since the battery status is not affected if the car and large vehicle version is used, it is important to ensure that the correct weighting for the battery status is allocated based on the specifications of the terminal type.

In order to evaluate the criteria, it is also useful to have some input from the user regarding the choice of segment. The input required from a user, for example, could be the importance of cost compared to the importance of QoS received. Other parameters could be the type of applications used and the choice of priority or prohibited segments. As an example, the user can be provided with the following interfaces, as illustrated in Figure 6.12, for their inputs.

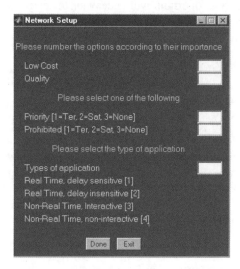

Figure 6.12   Example of user's preference input interface.

Ordinal Ranking

In fuzzy ordinal ranking, the segments are compared with each other to measure the performance of the segment in each criterion. In the fuzzification method, an expert-based system is employed in which a membership function of how each segment behaves in each criterion is first determined. Real measurements from the segments are then obtained and mapped onto the membership function to measure the performance of that particular segment.

By utilising fuzzy sets for the other six criteria, the ordinal ranking of the criteria to satisfy an objective for all the segments can be obtained. For example, if the criterion for charging model is used, it is important to determine the weighting in each segment in terms of cost. As an example, the GPRS segment could be assigned a higher ranking for being the cheapest, followed by the UMTS segment and the satellite segment.

The fuzzy ordinal ranking mechanism utilises the fuzzy ordering method described by [ROS-95], which uses a special notion of relativity outlined in [SHI-73]. Based on these pairwise comparisons, the segment that performs best in a particular criterion will be given a higher ranking.

Assuming that x and y are members of the available segments, a pairwise function, $f_y(x)$ and $f_x(y)$ can be defined, where:

$f_y(x)$ is a membership function of x with respect to y, and

$f_x(y)$ is a membership function of y with respect to x.

The relativity function is then given by the following equation:

$$f(x \mid y) = \frac{f_y(x)}{\max[f_y(x), f_x(y)]} \qquad (6.4)$$

The equation gives a measurement of the membership value of choosing x over y, where the $f(x|y)$ can be considered as a membership of preferring variable x over variable y. Using this equation, a square matrix of relativity values, C, can be constructed. This will be illustrated later. To find the overall ranking, it is suggested that the minimum values of each row in the matrix C is determined as the value that will have the lowest weight for ranking purposes. The minimum values in each row are then compared again to obtain the maximum values. The segment with the highest values has the highest priority. The advantage of using this method is that it allows a pairwise comparison between the alternatives rather than comparing all the different alternatives at the same time. Therefore, the alternatives are compared individually against their peers, which allows for a fairer and less biased comparison.

As an illustration, consider the charging model of the GMBS. First a pairwise membership function of the charging model is determined. This is shown below, where x_1, x_2 and x_3 represent GPRS, UMTS and Satellite, respectively. In order to perform the calculation, data on the cost of calls per unit for all the different segments is required. Assuming that the call rates are as follows:

- GPRS($cost\_x_1$) = 2/unit,

- UMTS($cost\_x_2$) = 3/unit,

- Satellite($cost\_x_3$) = 4/unit

The pairwise comparison could produce the following results. The objective is to give a higher weighting (near maximum value) when the objectives are fulfilled. The values given should be between 0 and 1.

$$
\begin{array}{lll}
f_{x1}(x_1) = 1 & f_{x1}(x_2) = 0.6 & f_{x1}(x_3) = 0.67 \\
f_{x2}(x_1) = 0.4 & f_{x2}(x_2) = 1 & f_{x2}(x_3) = 0.57 \\
f_{x3}(x_1) = 0.33 & f_{x3}(x_2) = 0.43 & f_{x3}(x_3) = 1
\end{array}
$$

Theoretically, the values given to the comparison can be arbitrarily assigned. However, for simplicity, the following equation is used to perform the pairwise comparison for the charging model. Different equations can be applied to different criteria.

$$f_{x_i}(x_j) = \frac{\cos t_- x_j}{\cos t_- x_i + \cos t_- x_j}.$$

(6.5)

Equation (6.5) can now be employed to calculate all the relativity values. The results are shown in the following matrix.

$$C = \begin{array}{c} \\ x_1 \\ x_2 \\ x_3 \end{array} \begin{array}{ccc} x_1 & x_2 & x_3 \\ \left[\begin{array}{ccc} 1.0000 & 1.0000 & 1.0000 \\ 0.6667 & 1.0000 & 1.0000 \\ 0.4925 & 0.7544 & 1.0000 \end{array} \right] \end{array}$$

From the C matrix, the minimum value in each row is determined, which results in the following matrix, C'.

$$C' = \begin{array}{c} \\ x_1 \\ x_2 \\ x_3 \end{array} \begin{array}{cccc} x_1 & x_2 & x_3 & min = f(x_i \,|\, X) \\ \left[\begin{array}{ccc} 1.0000 & 1.0000 & 1.0000 \\ 0.6667 & 1.0000 & 1.0000 \\ 0.4925 & 0.7544 & 1.0000 \end{array} \right] & \begin{array}{c} 1.0000 \\ 0.6667 \\ 0.4925 \end{array} \end{array}$$

From the matrix C', it can be seen that x_3, the GPRS is the cheapest segment, followed by the UMTS segment and the satellite segment. This can be determined by looking at the last column in matrix C'. GPRS has the highest values (1) compared to the other two segments.

The same approach can be adopted for signal strength, network latency, battery status, available bandwidth and reliability. However, a more robust method has to be devised in comparing criteria with real-time measurements for an accurate segment selection algorithm. This problem is more prominent for signal strength. One solution to this problem is to average the values received over certain predefined periods of time to smooth out the jitter in the signal. In addition, the fuzzy ordering procedure can be repeated a few times to obtain a more accurate result.

The ordinal ranking of priority segments is slightly different from the other criteria as this criterion is influenced by the input provided by the subscriber. Figure 6.12 shows a simple way of obtaining such information. Therefore, if the user does not indicate a priority segment, the segments are given the same values. However, if the subscriber chooses a priority segment, that segment will be given a higher ranking. The values given are arbitrary.

Fuzzification

When performing this step, data from the system is fed into a fuzzifier, to be converted into fuzzy sets. If the charging model is considered in a crisp set, the cost of using a segment can only be classed as expensive or inexpensive. If it is classed as expensive, it will be given a membership value of "1"; otherwise, it will have a membership value of "0". However, in a fuzzy set, the cost of a segment can be represented by anything between "0" and "1" depending on the membership function. The membership function is a curve or line that defines how each data or value is mapped onto a membership value and can be defined based

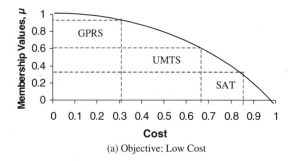

(a) Objective: Low Cost

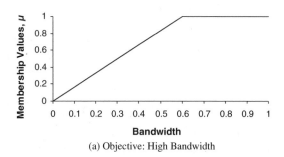

(a) Objective: High Bandwidth

Figure 6.13 Cost and bandwidth membership functions.

on the expert system methodology. In order to obtain the membership values for the fuzzy sets, the measurements for a particular parameter are mapped onto a membership function.

As an example, if the objective of the target segment is low cost, a membership function with the objective "low cost" can be constructed, as shown in Figure 6.13a. From the diagram, it can be inferred that if cost is low, the membership values will be high. Note that the x-values for cost are between "0" and "1". This could be modified later to include the actual cost of using each segment.

In the same diagram, the method by which the membership values are assigned can be demonstrated. If the cost obtained for UMTS is 0.5, the membership value for UMTS is 0.81. The same principle applies to all the three segments.

Several simulations were performed to test both methods of ordinal ranking in GMBS. From the results, it was found that the fuzzification method is more suitable for GMBS as it can adapt itself better to the different conditions. This approach will be adopted in the following sections.

Multiple Objective Decision Making

Once the ordinal ranking of the segments has been established, the second stage of the algorithm can be executed, where the importance of the objectives are rated based on the inputs of the user and the service providers.

Out of the nine criteria chosen for segment selection, only six criteria are in the fuzzy sets that will be used to determine the most suitable network. This is because two other criteria, segment availability and prohibited segments, are classed under crisp sets and have already been used to influence the ordinal ranking. For example, if a segment is unavailable or prohibited, it will be assigned zero values during the fuzzy ranking process, which results in the segment being ranked last. The remaining criterion, terminal type, will be used to influence the importance of the battery status. Therefore, only six criteria are involved in the final decision making; signal strength, charging models, network latency, reliability model, battery status and priority of segments. These criteria have to be graded in terms of their importance. Battery status is only important if the terminal type is a laptop computer. Otherwise, the terminal can be permanently attached to a vehicle.

The approach adopted for grading is based on the method described in [AGE-77], where the objectives are of varying degrees of importance, each with a numerical value assigned to it. This is obtained by determining the eigenvector of a matrix, which consists of paired comparison of the criteria. This numerical value, together with the values obtained from the fuzzy ordering procedure, is used to determine the most suitable segment.

In [AGE-77], a procedure, further developed by *Saaty* [SAA-80], is described for obtaining a ratio scale for a group of elements based upon the paired comparison of each element. By assuming that $C = \{C_1, \ldots, C_P\}$ is a set of p handover decision criteria, a set of values of relative importance can then be assigned to each criterion by performing a pairwise comparison between the designated criterion and the other criteria. For example, if C_1 is the designated criterion, then the set of values of relative importance for C_1 in comparison with the other criteria can be written as $\{a_{12}, a_{13}, \ldots, a_{1p}\}$ where $a_{ij, j \neq 1}$ is determined by comparing the importance of criterion C_1 to C_j. The system designer, based on the network capability, service requirements and the user's input, determines the values of a_{1j}. If C_i is more important than C_j, a_{ij} is assigned a number based on its degree of importance (Refer to Table 6.9). Then a matrix, B, of dimension $p \times p$ can be created where:

- $b_{ii} = 1$
- $b_{ij} = a_{ij}, i \neq j$
- $b_{ji} = 1/b_{ij}$.

Saaty also shows that by finding the unit eigenvector, W, corresponding to the maximum eigenvalue of B produces the cardinal ratio scale of the elements compared. To obtain the final weighting used in the decision process, α, W is multiplied with the number of criteria, n, resulting in the following matrix, E.

$$E = \begin{pmatrix} \alpha_1 \\ \alpha_2 \\ \alpha_3 \\ \alpha_n \end{pmatrix} = nW = \begin{pmatrix} nw_1 \\ nw_2 \\ nw_3 \\ nw_n \end{pmatrix}$$

Table 6.9 Definition for comparing importance of criteria [AGE-77]

Intensity of importance	Definition
1	Equal importance
3	Weak importance of one over the other
5	Strong importance of one over another
7	Demonstrated importance of one over the other
9	Absolute importance of one over the other
2, 4, 6, 8	Intermediate values; two adjacent judgements

To demonstrate how this procedure can be executed, some user inputs are necessary, as shown in Figure 6.12. The inputs that are relevant to this procedure is the ranking of the cost compared to the QoS perceived by the user and the types of application the user is currently using. These two inputs will be used to influence the importance of the charging model, bandwidth, network latency and reliability. The importance of battery status depends on the terminal type, whereas the importance of signal strength and priority of segments is at the discretion of the service provider or network designer.

For example, assuming that the following parameters are considered:

- Signal Strength is given highest priority. Therefore, when compared to other criteria, it is more important.

- Terminal Type used is portable. So, battery is given the second highest priority. When compared with the other parameters, it is also given a higher ranking, except when compared with signal strength.

- The user decides to choose cost over quality.

- Type of application is real-time and delay sensitive.

- The user decides to give more priority to the terrestrial segments.

- The network designer decides to give the same ratings to priority of segments when compared to cost and quality, respectively. This indicates that when a priority segment is compared to cost, it is given a rating of 1. The same approach is used when it is compared with QoS.

From the parameters above, matrix B can be constructed. The criteria of segment selection can be represented as C_1, C_2, C_3, C_4, C_5, C_6 and C_7, where $C_1 = $ Charging model, $C_2 = $ Bandwidth, $C_3 = $ Network Latency, $C_4 = $ Reliability, $C_5 = $ Signal Strength, $C_6 = $ Battery Status and $C_7 = $ Priority Segment. The resulting matrix is shown below. Note that the comparison is just an example and theoretically, any other values listed in Table 6.9 can be used to construct the matrix.

$$
B = \begin{array}{c} \\ C_1 \\ C_2 \\ C_3 \\ C_4 \\ C_5 \\ C_6 \\ C_7 \end{array}
\begin{array}{c} \begin{matrix} C_1 & C_2 & C_3 & C_4 & C_5 & C_6 & C_7 \end{matrix} \\
\begin{bmatrix}
1 & 3 & 3 & 3 & 1/3 & 1/3 & 1 \\
1/3 & 1 & 1 & 1 & 1/3 & 1/3 & 1 \\
1/3 & 1 & 1 & 1 & 1/3 & 1/3 & 1 \\
1/3 & 1 & 1 & 1 & 1/3 & 1/3 & 1 \\
3 & 3 & 3 & 3 & 1 & 3 & 3 \\
3 & 3 & 3 & 3 & 1/3 & 1 & 3 \\
1 & 1 & 1 & 1 & 1/3 & 1/3 & 1
\end{bmatrix}
\end{array}
$$

Since signal strength is classed as having the most important priority, it is given an importance rating of 3. Note that the comparison is pairwise. Since cost is given more priority than quality, cost is given an importance rating of 3 when compared to network latency, available bandwidth and reliability. Furthermore, since the user is utilising real-time and delay sensitive applications, the network latency, bandwidth and reliability are given the same importance rating when compared with each other.

However, if the user changes the application from real-time, delay sensitive to non-real-time, and interactive, the following matrix B could be obtained. This is because, high bandwidth might not be so important compared to network latency and reliability in such cases.

$$B = \begin{array}{c} \\ C_1 \\ C_2 \\ C_3 \\ C_4 \\ C_5 \\ C_6 \\ C_7 \end{array} \begin{array}{c} \begin{array}{ccccccc} C_1 & C_2 & C_3 & C_4 & C_5 & C_6 & C_7 \end{array} \\ \left[\begin{array}{ccccccc} 1 & 3 & 3 & 3 & 1/3 & 1/3 & 1 \\ 1/3 & 1 & 1/3 & 1/3 & 1/3 & 1/3 & 1 \\ 1/3 & 3 & 1 & 1 & 1/3 & 1/3 & 1 \\ 1/3 & 3 & 1 & 1 & 1/3 & 1/3 & 1 \\ 3 & 3 & 3 & 3 & 1 & 3 & 3 \\ 3 & 3 & 3 & 3 & 1/3 & 1 & 3 \\ 1 & 1 & 1 & 1 & 1/3 & 1/3 & 1 \end{array} \right] \end{array}$$

From the matrix, the eigenvector corresponding to the matrix can be calculated. The results of the eigenvectors for the first matrix are shown in the following matrix, W.

$$W = \begin{pmatrix} 0.3284 \\ 0.1276 \\ 0.1959 \\ 0.1959 \\ 0.7083 \\ 0.5108 \\ 0.1911 \end{pmatrix}$$

Therefore, matrix $E = nW$, where n in this case is 7. The matrix, E, is as follows:

$$E = \begin{pmatrix} 2.2988 \\ 0.8932 \\ 1.3711 \\ 1.3711 \\ 4.9582 \\ 3.5756 \\ 1.3375 \end{pmatrix}$$

If the objective of the system is associated with a fuzzy subset, the decision D becomes

$$D = C_1 \cap C_2 \cap C_3 \cap C_4 \ldots$$

Then, by using the values found from matrix E, α, and raising the fuzzy set to α, the optimum segment can be found where the decision D, in terms of fuzzy subsets becomes:

$$D = C_1^{\alpha_1} \cap C_2^{\alpha_2} \cap C_3^{\alpha_3} \cap C_4^{\alpha_4} \ldots \cap C_n^{\alpha_n} \qquad (6.6)$$

As an example to illustrate the fuzzy logic based segment selection procedure, assuming that all operations required for the ordinal ranking are complete and the fuzzy ordering of all the criteria are as follows:

$$Cost = \begin{bmatrix} 1.0000 \\ 0.6667 \\ 0.4925 \end{bmatrix} \quad Bandwidth = \begin{bmatrix} 0.0753 \\ 1.0000 \\ 0.0753 \end{bmatrix} \quad Network\ Latency = \begin{bmatrix} 0.6667 \\ 1.0000 \\ 0.1111 \end{bmatrix} \quad Reliability = \begin{bmatrix} 0.4286 \\ 1.0000 \\ 0.2500 \end{bmatrix}$$

$$Signal\ Strength = \begin{bmatrix} 0.6667 \\ 1.0000 \\ 0.5714 \end{bmatrix} \quad Battery\ Status = \begin{bmatrix} 0.4286 \\ 1.0000 \\ 0.2500 \end{bmatrix} \quad Priority\ Segment = \begin{bmatrix} 0.6600 \\ 0.6600 \\ 0.3300 \end{bmatrix}$$

If these seven criteria are then fed into the multiple objective decision making algorithm, as C_1, C_2, C_3 and so on, respectively, it is then possible to choose the most suitable segment. The column on the left shows the values of the fuzzy set before the eigenvector values are applied to it to emphasise the importance of certain objectives. The column on the right shows the results after the importance ratings are applied.

$$E = \begin{pmatrix} 2.2988 \\ 0.8932 \\ 1.3711 \\ 1.3711 \\ 4.9582 \\ 3.5756 \\ 1.3375 \end{pmatrix}$$

Before	After

$$C_1 = \left\{ \frac{1.0000}{GPRS}, \frac{0.6667}{UMTS}, \frac{0.4925}{SAT} \right\} \quad C_1^{\alpha_1} = \left\{ \frac{1.0000}{GPRS}, \frac{0.3938}{UMTS}, \frac{0.1963}{SAT} \right\}$$

$$C_2 = \left\{ \frac{0.0753}{GPRS}, \frac{1.0000}{UMTS}, \frac{0.0753}{SAT} \right\} \quad C_2^{\alpha_2} = \left\{ \frac{0.0993}{GPRS}, \frac{1.0000}{UMTS}, \frac{0.0993}{SAT} \right\}$$

$$C_3 = \left\{ \frac{0.6667}{GPRS}, \frac{1.0000}{UMTS}, \frac{0.1111}{SAT} \right\} \quad C_3^{\alpha_3} = \left\{ \frac{0.5736}{GPRS}, \frac{1.0000}{UMTS}, \frac{0.0492}{SAT} \right\}$$

$$C_4 = \left\{ \frac{0.4286}{GPRS}, \frac{1.0000}{UMTS}, \frac{0.2500}{SAT} \right\} \quad C_4^{\alpha_4} = \left\{ \frac{0.3130}{GPRS}, \frac{1.0000}{UMTS}, \frac{0.1495}{SAT} \right\}$$

$$C_5 = \left\{ \frac{0.6667}{GPRS}, \frac{1.0000}{UMTS}, \frac{0.5714}{SAT} \right\} \quad C_5^{\alpha_5} = \left\{ \frac{0.1340}{GPRS}, \frac{1.0000}{UMTS}, \frac{0.0624}{SAT} \right\}$$

$$C_6 = \left\{ \frac{0.4286}{GPRS}, \frac{1.0000}{UMTS}, \frac{0.2500}{SAT} \right\} \quad C_6^{\alpha_6} = \left\{ \frac{0.0483}{GPRS}, \frac{1.0000}{UMTS}, \frac{0.0070}{SAT} \right\}$$

$$C_7 = \left\{ \frac{0.6600}{GPRS}, \frac{0.6600}{UMTS}, \frac{0.3300}{SAT} \right\} \quad C_7^{\alpha_7} = \left\{ \frac{0.5736}{GPRS}, \frac{0.5736}{UMTS}, \frac{0.2270}{SAT} \right\}$$

Then, by applying Eq. (6.6) onto the fuzzy set, the following decision model is obtained.

$$D = \left\{ \frac{0.0483}{GPRS}, \frac{0.3938}{UMTS}, \frac{0.0070}{SAT} \right\}$$

From the decision model, the selected segment is UMTS, as it has the highest values compared to the other two segments.

6.5.4.6 Analysis of Algorithm

Handover Initiation

Simulation Parameters

In handover initiation, the main objective is to demonstrate how the value of the handover factor changes when the inputs of the measurements obtained from the network are varied using a normal distribution. This is performed to ensure that handover is initiated correctly.

The simulation model is shown in Figure 6.14. Three scenarios are investigated, to represent the possible measurements obtained from each segment. Four inputs are fed into the fuzzy logic controller, their attributes are shown in Table 6.10.

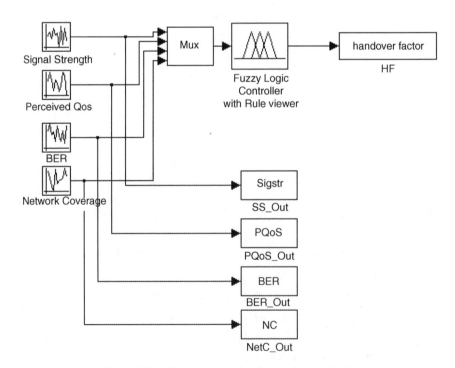

Figure 6.14 Simulation model for handover initiation.

Table 6.10   Simulation parameters for handover initiation

Criteria	Scenario 1		Scenario 2		Scenario 3	
	Mean	Var	Mean	Var	Mean	Var
Signal Strength	0.2	0.2	0.5	0.2	0.8	·0.2
QoS	0.3	0.2	0.5	0.2	0.7	0.2
BER	0.6	0.2	0.5	0.2	0.4	0.2
Network Coverage	0.3	0.2	0.5	0.2	0.7	0.2

Results and Analysis

The cumulative distribution function (cdf) for the handover factor is shown in Figure 6.15. Since the values fed into the fuzzy logic controller in scenario 1 borders on the weak side, the values obtained for the handover factor are clustered around 0.7 to 0.8. This indicates that handover should be initiated. In contrast, the inputs fed to the controller in scenario 3 are quite strong. As a result, the handover factor is low and handover is not initiated. Thus, it can be concluded that the algorithm behaves correctly and produces correct handover responses under different operating environments.

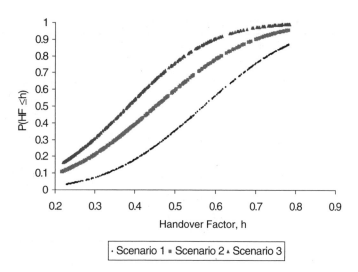

Figure 6.15   Cumulative distribution function for handover factor.

Segment Selection and Handover Decision

Since the same algorithm is used for segment selection and handover decision, the simulation parameters for both will also be the same. The objectives of this simulation are as follows:

- To investigate the behaviour of the segment selection and handover decision algorithm. The simulation aims to demonstrate that by changing certain inputs into the system, the choice of the selected segment will differ. This is performed by varying the application used by the user and also the measurements received by the segment.

- To perform a sensitivity analysis on the criteria used in the simulation. The results of this analysis are useful since they provide an indication of how certain criteria or network characteristics for a particular segment can be improved to make them on par with other wireless segments.

Simulation Parameters

a) Behavioural analysis

Two sets of simulations are analysed here. The first set is used to investigate how the weighting matrix is affected by the applications used, whereas the second set of results investigate how the measurements obtained by the terminal affect the result of segment selection. In both sets, the preference of the user will alternate between cost (case a) and QoS (case b). The simulation parameters are shown in Table 6.11. The following data rates are applied to the system: UMTS = 384 kbps, GPRS = 150 kbps and M-ESW = 160 kbps. In the network attributes, each segment is assigned values ranging from 0 to 1. A higher rating indicates a higher ranking.

Table 6.11   Simulation parameters for behavioural analysis

	Criteria		Attributes		
	Applications		Real-time delay sensitive (A)		
Simulation 1			Real-time, delay insensitive (B)		
			Non-real-time, interactive (C)		
			Non-real-time, non-interactive (D)		
	Network attributes		GPRS	UMTS	SAT
	Cost	charging	0.90	0.80	0.70
		Bandwidth	0.80	0.90	0.70
Simulation 2	QoS	Latency	0.60	0.80	0.50
		Reliability	0.90	0.90	0.50
		Signal strength	0.70	0.69	0.80
	Others	Battery	0.90	0.60	0.70
		Segment priority	0.50	0.50	0.50

b) Sensitivity analysis

As mentioned previously, the main objective of the simulation is to perform a sensitivity analysis on the criteria used in the simulation. The results of this analysis are useful since they provide an indication of how certain criteria or network characteristics for a particular segment can be improved to make them on par with the other wireless segments. In this analysis, two sets of simulations are also performed.

Table 6.12 Simulation parameters for sensitivity analysis

Criteria	Weightings α	Measurements of criteria, C		
		GPRS	UMTS	SAT
Simulation 1				
Charging model	2.6458	0.9	0.6	0.3
Bandwidth	2.6458	Varies	Varies	Varies
Latency	2.6458	0.9	0.9	0.9
Availability	2.6458	0.9	0.9	0.9
Signal strength	2.6458	0.9	0.9	0.9
Battery	2.6458	0.9	0.9	0.9
Segment priority	2.6458	0.9	0.9	0.9
Simulation 2				
Charging model	1.9589	0.9	0.6	0.3
Bandwidth	3.9926	Varies	Varies	Varies
Latency	2.4175	0.9	0.9	0.9
Availability	2.4175	0.9	0.9	0.9
Signal strength	2.4175	0.9	0.9	0.9
Battery	2.4175	0.9	0.9	0.9
Segment priority	2.4175	0.9	0.9	0.9

Since the emphasis of the simulation is to vary the characteristics of the network parameters, all weightings for the criteria will initially be the same. The values for α shown in Table 6.12 in Simulation 1 are obtained by providing equal importance to all the seven criteria, whereas in Simulation 2, bandwidth is given a demonstrated importance over cost. Out of the seven criteria specified for the algorithm, only two criteria will be varied: the charging model and the available bandwidth, with the simulation parameters shown in Table 6.12. The first simulation is a controlled simulation to show the relationship between cost and bandwidth. In the simulation, it is assumed that the cost of using the GPRS segment is cheaper than the other two segments. In the second simulation, the weightings are changed to show the effect of the weighting factor on the algorithm. It is also assumed that the application is tailored to suit a business user. Therefore, bandwidth is given a higher weighting compared to cost.

Results and Analysis

a) Behavioural analysis

- Simulation 1

The first set of results for segment selection is shown in Figure 6.16. From the diagram, it can be seen that user requirements play an important role in determining the weighting for each criterion. When the user prefers cost to QoS, the charging model is given higher weighting. In contrast, the QoS parameters are given a higher weighting in case (b). When

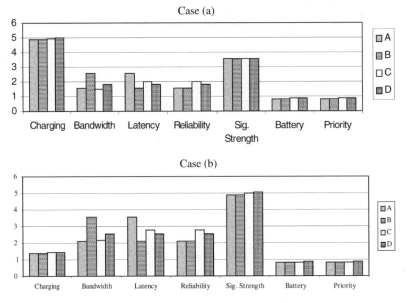

Figure 6.16   Comparison of the weighting matrix.

comparing QoS parameters, signal strength is given higher weighting as it is considered more important than the other three QoS parameters. Meanwhile, the types of user application influence the three other QoS parameters. The weightings are varied depending on the applications used.

- Simulation 2

   Table 6.13 shows the second set of results when the values in Table 6.11 are assigned to the network. In Case (a), even though the signal strength for UMTS is lower than that of the satellite segment, it is chosen for real-time and delay sensitive applications because the user prefers cost to quality. For non-real-time services, GPRS is chosen since it is the cheapest segment when compared to the other two segments. In case (b), the user decides to select quality over cost, therefore, different results are obtained. The results depend primarily on

Table 6.13   Result for simulation 1 (Sensitivity Analysis)

Selected segment	Conditions
GPRS	C^α (Bandwidth_GPRS) $\geq C^\alpha$ (Bandwidth_UMTS) AND C^α (Bandwidth_GPRS) $\geq C^\alpha$ (Bandwidth_SAT)
UMTS	C^α(Bandwidth_UMTS) $\geq C^\alpha$ (Bandwidth_GPRS) AND C^α(Cost_UMTS) $\geq C^\alpha$ (Bandwidth_GPRS) AND C^α(Bandwidth_UMTS) $\geq C^\alpha$(Bandwidth_SAT)
SAT	C^α (Bandwidth_SAT) $\geq C^\alpha$ (Bandwidth_GPRS) AND C^α(Cost_SAT) $\geq C^\alpha$ (Bandwidth_GPRS) AND C^α(Bandwidth_SAT) $\geq C^\alpha$ (Bandwidth_UMTS) AND C^α(Cost_SAT) $\geq C^\alpha$ (Bandwidth_UMTS)

the applications used. When real-time and delay sensitive applications are selected, the chosen segment is UMTS because of its lower latency. This changes when delay insensitive applications are used. In such cases, latency is not so important and since the satellite segment has a higher signal strength, it is chosen over UMTS. For non-real-time and interactive services, GPRS is chosen over UMTS because it has a better signal strength. When compared to the satellite segment, GPRS is still chosen although it has weaker signal strength as the satellite segment is weaker when other criteria are taken into account. These results show that the algorithm can adapt itself to satisfy the requirements of the user.

b) Sensitivity analysis

- Simulation 1

Table 6.14 generally provides the conditions that have to be fulfilled by each segment when the charging model defined in Table 6.12 is implemented on the system. From the three segments chosen in the simulation, it can be seen from Table 6.12 that the satellite segment is the least favourable when mapped onto the membership function of "low cost". Therefore, if the network provider of the satellite segment wishes to make the segment more desirable, the bandwidth of the satellite has to be set higher than the bandwidth of the other two segments. The required bandwidth depends on the membership function defined in Figure 6.13. However, in addition to the bandwidth, the membership values of bandwidth for the UMTS and GPRS segments must still be set lower than the membership values of cost for SAT. This is because the algorithm to obtain D uses the fuzzy intersection rule and since no weightings are applied to the algorithm in Simulation 1, the minimum membership values in each segment act as the constraining factor.

Table 6.14   Result for segment selection

Conditions		Result
Case (a) Cost	Real-time, delay sensitive	UMTS, SAT
	Real-time, delay insensitive	UMTS, SAT
	Non-real-time, interactive	GPRS, UMTS, SAT
	Non-real-time, non-interactive	GPRS, UMTS, SAT
Case (b) Quality	Real-time delay sensitive	UMTS, SAT
	Real-time, delay insensitive	SAT, UMTS
	Non-real-time, interactive	GPRS, UMTS, SAT
	Non-real-time, non-interactive	SAT, GPRS, UMTS

- Simulation 2

The main objective of Simulation 2 is to demonstrate that by changing the weighting applied to each criterion, it will be easier to consider the trade-off between cost and bandwidth. The results obtained are shown in Table 6.15. For the purpose of this simulation, the results shown are aimed at selecting the satellite segment as the most suitable segment since this segment is considered the most undesirable when cost is first taken into consideration.

Table 6.15   Result for simulation 2 (Sensitivity Analysis)

	Without weighting			With weighting		
	Cost (C^α)	Bandwidth		Cost (C^α)	Bandwidth	
		Required C^α	Required C		Req. C^α	Req. C
GPRS	0.90	0.30	0.63	0.81	0.09	0.55
UMTS	0.60	0.30	0.63	0.37	0.09	0.55
SAT	0.30	0.31	0.64	0.09	0.09	0.56

From the results, it can be observed that when a weighting system is implemented in the system, the required bandwidth for satellite selection is reduced and hence the influence of cost is decreased. This approach, therefore, eases the process of trading off the bandwidth and cost.

REFERENCES

[AGE-77] R.R. Ager: Multiple Objective Decision Making using Fuzzy Sets, *International Journal of Man-Machine Studies*, **9**(4), 1977; 375–382.

[3GPP-99] 3GTS22.115 v3.2.0 (1999-10) Service Aspects; Charging and Billing.

[3GPP-00] 3G TS 22.129 v3.4.0 (2000-10): Technical Specification Group Services and Systems Aspects: Service Aspects; Handover Requirements between UMTS and GSM or other Radio Systems.

[CHA-99] D. Chalmers, M. Sloman: A Survey of Quality of Service in Mobile Computing Environments, *IEEE Communications Surveys*, **2**(2), Second Quarter 1999; 2–10.

[CON-00] P. Conforto, C. Tocci, G. Losquadro, R.E. Sheriff, P.M.L. Chan: Global Mobility and QoS Provision for Internet Services: the SUITED Solution, *Proceedings of IEEE Globecom 2000 Workshop CFP: Service Portability and Virtual Customer Environments*, San Francisco, 27 November–1 December 2000; 3–12.

[DAN-00] M.S. Dang, A. Prakash, D.K. Anvekar, D. Kapoor, R. Shorey: Fuzzy Logic Based Handoff in Wireless Networks, *Proceedings of the 51st Vehicular Technology Conference (VTC 2000 Spring)*, Tokyo, 15–18 May 2000; 2375–2379.

[EDW-94] G. Edwards, R. Shanker: A New Hand-Off Algorithm Using Fuzzy Logic, *Proceedings of Southeastcon'94: Creative Technology Transfer—A Global Affair*, 10–13 April 1994; 89–92.

[EDW-00] G. Edwards, A. Kandel, R. Shanker: Fuzzy Handoff Algorithms for Wireless Communications, *Fuzzy Sets and Systems*, **110**(3), March 2000; 379–388.

[EFT-98] N. Efthymiou, Y.F. Hu, R.E. Sheriff, A. Properzi: Inter-Segment Handover Algorithm for an Integrated Terrestrial/Satellite Environment, *Proceedings of 9th IEEE International Symposium on Personal, Indoor and Mobile Radio Communications (PIMRC)*, Boston, Massachusetts, USA, 8–11 September 1998; 993–998.

[ETS-02a] ETSI TS 101 344 V7.9.0 (2002-09) Digitial Cellular Telelcommunications System (Phase 2+); General Packet Radio Service (GPRS), Service Description; Stage 2 (3GPP TS 03.60 version 7.9.0 Release 1998).

[ETS-02b] ETSI TS 123 060 V5.2.0 (2002-06) Digitial Cellular Telelcommunications System (Phase 2+); Universal Mobile Telecommunications Systems (UMTS); General Packet Radio Service (GPRS) Service Description; Stage 2 (3GPP TS 23.060 version 5.2.0 Release 5).

[ETS-02c] ETSI TS 125 323 V5.2.0 (2002-09) Universal Mobile Telecommunications System (UMTS); Packet Data Convergence Protocol (PDCP) Specification (3GPP TS 25.323 version 5.2.0 Release 5).

[GHO-98] S. Ghosh: A Survey of Recent Advances in Fuzzy Logic in Telecommunications Networks and New Challenges, *IEEE Transactions on Fuzzy Systems*, **6**(3), August 1998; 443–447.

[IET-02] C. Perkins (Ed.): IP Mobility Support for IPv4, *Internet Engineering Task Force* RFC 3220, January 2002.

[IET-03a] D.B. Johnson, C. Perkins, J. Arkko: Mobility Support in IPv6, *Internet Engineering Task Force*, Internet Draft draft-ietf-mobileip-ipv6-24.txt, Work in Progress, 30 June 2003.

[IET-03b] R. Droms *et al.*: Dynamic Host Configuration Protocol for IPv6 (DHCPv6), *Internet Engineering Task Force*, RFC 3315, July 2003.

[IET-97] R. Droms: Dynamic Host Configuration Protocol, *Internet Engineering Task Force*, RFC 2131, March 1997.

[LAU-95] S.S. Lau, K. Cheung, J.C. Chuang: Fuzzy Logic Adaptive Handoff Algorithm, *Proceedings of IEEE Globecom'95*, Singapore, 13–17 November 1995; 504–508.

[MOU-92] M. Mouly, M. Pautet: The GSM System for Mobile Communications, Palaisieu, France, 1992.

[MUR-91] A. Murase, I.C. Symington, E. Green: Handover Criterion for Macro and Microcellular Systems, *Proceedings of the IEEE Vehicular Technology Conference (VTC'91)*, St. Louis, MO, USA, 19 May 1991; 524–530.

[POL-96] G.P. Pollini: Trends in Handover Design, *IEEE Communications Magazine*, **34**(3), March 1996; 82–90.

[ROS-95] T. J. Ross: Fuzzy Logic with Engineering Applications, *McGraw-Hill*, Inc., USA, 1995.

[SAA-80] T.L. Saaty: The Analytical Hierarchy Process, *McGraw-Hill*, Inc., USA, 1980.

[SHE-01] R.E. Sheriff, Y.F. Hu: Mobile Satellite Communication Networks, *Wiley*, 2001.

[SHI-73] M. Shimura: Fuzzy Sets Concept in Rank-Ordering Objects, *Journal of Mathematical Analysis and Applications*, **43**(3), 1973; 717–733.

[TAB-00] S. Tabbane: Handbook of Mobile Radio Network; *Artech House*, 2000.

[WAN-99] H.J. Wang, R.H. Katz, J. Giese: Policy Enabled Handoffs in Heterogeneous Wireless Networks, *Proceedings of 2nd IEEE Workshop on Mobile Computing and Applications (WMCSA '99)*, New Orleans, USA, 25–26 February 1999; 51–60.

ACRONYMS

2G	Second-Generation
3G	Third-Generation
AS	Autonomous System
BB	Bandwidth Broker
BER	Bit Error Rate
BGP	Border Gateway Protocol
BTS	Base Transceiver Station
CoA	Care-of-Address
DHCP	Dynamic Host Configuration Protocol
FA	Foreign Agent
F-ISP	Federated ISP
FES	Fixed Earth Station
GGSN	Gateway GPRS Support Node
GMBS	Global Mobile Broadband System
GMMT	GMBS Multi-Mode Terminal
GPRS	General Packet Radio Service
GPS	Global Positioning System
GRIP	Gauge&Gate Reservation with Independent Probing
GSM	Global System for Mobile Communications
GSN	GPRS Support Node

GSP	GMBS Service Provider
GTP	GPRS Tunnelling Protocol
HA	Home Agent
HPLMN	Home PLMN
IMSI	International Mobile Subscriber Identity
IP	Internet Protocol
ISHO	Inter-segment Handover
ISP	Internet Service Provider
IWU	Inter-Working Unit
LAN	Local Area Network
MAHO	Mobile Assisted Handover
MCHO	Mobile Controlled Handover
MHHA	Mobile Host Home Agent
MM	Mobility Management
MRHA	Mobile Router Home Agent
MS	Mobile Station
MT	Mobile Terminal
NAHO	Network Assisted Handover
NAP	Network Access Point
NC	Network Coverage
NLAPI	Network Layer Access Point Identifier
NCHO	Network Controlled Handover
NOC	Network Operation Centre
OLN	Overlaying Network
PDN	Packet Data Network
PDP	Packet Data Protocol
PDU	Packet Data Unit
PLMN	Public Land Mobile Network
PS	Packet Switched
QASM	QoS Support Module
QoS	Quality of Service
RNC	Radio Network Controller
RRC	Radio Resource Control
SaT	Satellite Terminal
SGSN	Serving GPRS Support Node
TCP	Transmission Control Protocol
TE	Terminal Equipment
TID	Tunnel Identifier
T-IWU	Terminal Inter-Working Unit
TLLI	Temporary Logical Link Layer Identity
UDP	User Datagram Protocol
UMTS	Universal Mobile Telecommunications System
URA	UTRAN Registration Area
UTRA	UMTS Terrestrial Radio Access
UTRAN	UTRA Network
VPLMN	Visited Public Land Mobile Network
W-LAN	Wireless LAN

7

Network Protocol Design

PAULINE M.L. CHAN[1], PAOLO CONFORTO[2], Y. FUN HU[1] and
RAY E. SHERIFF[1]

[1]University of Bradford, UK; [2]Alenia Spazio, Italy

7.1 INTRODUCTION

Network protocols define how entities within a network communicate with each other. This
is achieved through the exchange of structured messages in a specified sequence over time.
For example, such messages may contain specific information to be acted on or can simply
serve the purpose of acknowledging a communication. Network protocols are not restricted
to a particular environment or application. They can involve communication between entities
or modules within a single item of equipment, or between items of equipment within a
network or between disparate networks to form a single heterogeneous environment, such as
in the Global Mobile Broadband System (GMBS). The latter example is known as inter-
working. This involves the definition of protocols that allow otherwise incompatible
networks to communicate with each other.

The means of designing such protocols is becoming increasingly complex and as a
consequence, the need to employ some form of design methodology is required.

A "design methodology" can be thought of as a method that assists in the design process,
starting from the user or functional requirements and finishing at the real system implementation. The methodology presented in the following is mainly based on the initial
concepts designed by the EC-funded RACE II project MoNet, the ITU (International
Telecommunication Union) Recommendations I.130 [ITU-98] and Q.65 [ITU-00] and the
Intelligent Network Conceptual Model. MoNet, a pan-European project performed during
the mid-90s, considered the network requirements of the UMTS (Universal Mobile
Telecommunications System). ITU Recommendation I.130 defines the method for describing switching and signalling for ISDN (Integrated Services Digital Network). ITU Recommendation Q.65 presents the Unified Functional Methodology (UFM) by extending the
concepts described in [ITU-98] to include services provided by networks of different types
such as UMTS and B-ISDN (Broadband ISDN). The UFM provides two alternative means
for developing functional descriptions of services. In the following, the process of specifying
functional entities (FEs), information flows (IFs) and the use of the Specification and
Description Language (SDL) will be presented. The application of SDL is defined in the ITU

Space/Terrestrial Mobile Networks. Edited by R.E. Sheriff, Y.F. Hu, G. Losquadro, P. Conforto, C. Tocci
© 2004 John Wiley & Sons, Ltd ISBN: 0-470-85031-0

Recommendations Z.100 [ITU-99c] and Z.120 [ITU-99d]. The alternative approach provided by the ITU in Q.65 follows an object-oriented technique.

In each step of the design process, only a subset of functionalities is taken into account. These functionalities become more specific as the development progresses and the subject of the design process becomes more elaborated. Each step acts as a starting point for the next step, where another set of functionalities is then considered.

The overall GMBS network comprises of a substantial number of network functionalities involving many different network components. It would be impractical to present within the confines of a Chapter the complete protocol design of the network. However, as the design process is generic, it is possible to present the methodology using as an example a single, relatively straightforward network functionality. With this in mind, the following sections concentrate on the GMBS Registration procedure. This involves generic functionality, as well as functionality that is specific to a particular access segment, that is UMTS, GPRS (General Packet Radio Service), W-LAN (Wireless Local Area Network) or Mobile-EuroSkyWay (M-ESW). In the following example, issues specific to the satellite access network are considered.

7.2 FUNCTIONAL MODELS, INFORMATION FLOWS AND FUNCTIONAL ENTITIES ACTIONS

7.2.1 Introduction

Six steps are essentially required for the characterisation of services and network capabilities, as defined in [ITU-00]. These include the definition of the "functional model" (FM), "Service Independent Building Blocks" (SIB), "Information Flow" (IF) diagrams, "Specification and Description Language" diagrams and the mapping of the entities into physical locations. The definition of SIB and SDL is optional in the specifications. The FM, which is the first major step in the design process, can be derived from the functional requirements of the network. The FM comprises of "Functional Entities" (FEs) and the relations between them are specified by means of IFs.

An FE is a set of service providing functions in a single spatial location that cannot be divided any further when mapped onto a physical entity (PE). An FE is also a subset of the total set of functionalities needed to provide a service.

In the IF diagram, several notations are used to describe the overall process. These are listed as follows:

1. Only lines that go through a certain FE will be processed by the FE.

2. If a Functional Entity Action (FEA) (labelled by number) is placed below a line when the FE receives a message, the action is dependent on the receipt of the message. This indicates that the action can only be performed after receiving a signal.

3. Similarly, if an FEA is placed above a line when the FE is sending a message, the action must be executed before the message can be sent and *vice-versa*.

4. Basic call flows are shown by dashed lines and chevrons. Supplementary service flows are shown by solid lines and arrow heads.

An FEA contains descriptions of actions required for each FE and is identified by a reference number. Each number is of the form *XYZ*, where X describes the type of service action. For supplementary service action, X of the FEA number must be a "9". Y is the number of the FE where the action is executed, whereas Z essentially enumerates the actions in a single FE ($Z = 1, \ldots, 9, A, \ldots, Z, a, \ldots, z$).

By way of an example, and as noted above, FMs, IFs and FEAs are defined for the GMBS Registration procedure. The GMBS Registration procedure, like many of the procedures identified for GMBS, involves communication between entities that are applicable to all segment specific entities, termed generic procedures, and the communication between entities that are specific to each access segment. The communication between the GMBS generic procedures and the segment specific procedures define how the respective networks inter-work to form a common GMBS architecture.

7.2.2 GMBS Registration

7.2.2.1 Generic GMBS Procedures

Functional Model

GMBS Registration refers to procedures that are executed when the GMBS Multi-Mode Terminal (GMMT) is switched-on, such as obtaining access to the network, service provider selection and authentication. When the GMMT is switched-on, segment specific registration would be performed before the Terminal Inter-Working Unit (T-IWU) is made aware of segment availability.

The FEs required for GMBS Registration are shown in Figure 7.1.

Figure 7.1 Functional model for GMBS registration.

This model can be seen to comprise of three FEs, the convention being to represent an FE by a circular enclosure with the name of the FE within it. It should be noted, however, that each access segment would have its own FE associated with the Service Handler operation. This is indicated in the above figure through the use of a dotted-line Service Handler FE partially hidden behind the generic representation, shown with a solid line. From the perspective of the RHT entity, each Service Handler will behave in a similar way, however, as far as the access networks are concerned, each will function specifically to meet the operations of the particular access segment. Each FE performs a specific task. A line between FEs indicates that there is a communication between them. Thus, in Figure 7.1, it can be concluded that the entity RHT communicates with the Service Handler entity and also with the Current Location Information (CLI) entity. It can also be concluded that there is no direct communication between the Service Handler and CLI entities. The functionality of each entity is as follows:

Registration Handler-Terminal (RHT). This entity receives the segment availability information from the segment specific mobile terminal (SS-MT) and then passes this information to the FE responsible for updating information on segment availability. It is also responsible for registering users with the GMMT.

Current Location Information. This entity is responsible for receiving information on the availability of each segment from the RHT entity.

Service Handler. Service Handler represents the generic representation of the service handlers that are present in each specific segment (i.e. UMTS, M-ESW, GPRS, W-LAN). How this performs in the satellite network will be considered shortly, when segment specific functionalities are integrated with those of GMBS, using the satellite segment as an example.

The mechanisms associated with communicating between entities are defined by *Information Flows.*

Information Flow

Two procedures will be initiated when the GMMT is switched-on. The individual SS-MTs perform segment specific registration. At the same time, the RHT entity requests information on segment availability from each segment before waiting for a response from each SS-MT Service Handler. This is shown in Figure 7.2 and detailed in Table 7.1. When the RHT entity

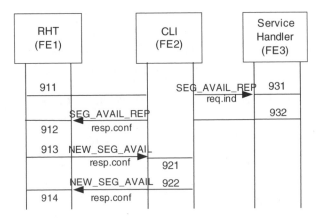

Figure 7.2 Information flow for GMBS registration.

Table 7.1 Messages exchanged during GMBS registration

Relationship	Item	req.ind	resp.conf
SEG_AVAIL_REP			
r_1	Request for segment availability following initialisation	Mandatory	
r_1	Segment availability following segment specific registration		Mandatory
NEW_SEG_AVAIL			
r_2	Information of new segments	Mandatory	
r_2	Confirmation of the information of new segments		Optional

Table 7.2   FEAs during GMBS registration

Functional entities	FEAs	Function
RHT (FE1)	911	Initialisation of the T-IWU.
		Creates a request to probe the availability of each segment.
	912	Receives segment availability report from Service Handlers.
	913	Checks a local database for availability of new segment. If new segment exists, a new signal is created to inform CLI of the new segment.
	914	Receives confirmation of the Registration procedure.
CLI (FE2)	921	Receives information of new segment from RHT.
		Updates the local database with information of the new segment.
	922	Creates a signal to RHT after updating the information of the new segment.
Service Handler (FE3)	931	Performs segment specific registration upon start up.
		Receives request for segment availability from RHT.
	932	Upon successful registration, the Service Handler will send information of segment registration to RHT.

receives registration confirmation and segment availability information from the SS-MTs, it passes this information to the CLI entity so that location update procedures can be performed before segment selection.

Functional Entity Actions

Table 7.2 shows the definition of the FEAs for the Registration procedure. The table shows the different actions or functionalities executed by the FE before and after the FE receives the signalling messages.

7.2.2.2   M-ESW Segment Specific Procedures

Functional Model

This section integrates the M-ESW FM into the overall GMBS FM. In particular, a description of both the M-ESW FEs in charge of interacting with the GMBS FEs, and the M-ESW FEs involved in the M-ESW message IFs triggered in order to execute GMBS procedures are presented.

Two FEs specific to the satellite segment can be identified from the FM in Figure 7.3, namely:

Figure 7.3   Functional model for M-ESW service handler.

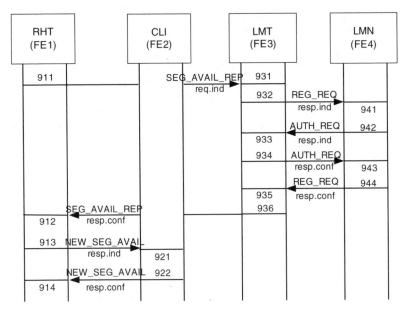

Figure 7.4   Signalling flow during GMBS registration.

Location Management-Terminal (LMT): This handles, by co-operating with the corresponding FE at the network side, the M-ESW Registration procedure. It is located in the broadband satellite mobile terminal (SAT-MT).

Location Management-Network (LMN): This is located in the M-ESW Network Operation Centre (NOC), and handles the M-ESW Registration procedure.

Information Flow

When the GMMT is switched-on, the SAT-MT, upon completing the synchronisation procedure, requires a Registration/Authentication service from the M-ESW network, providing log-on functionality for user access management.

In the IFs depicted in Figure 7.4, the M-ESW specific messages are identified by the signalling exchanges between FE3 and FE4 (LMT and LMN).

The functionality of the SAT messages is described in Table 7.3.

Table 7.3   M-ESW messages exchanged during GMBS registration

Relationship	Item	req.ind	resp.conf
REG_REQ			
r_3	Registration Request	Mandatory	
r_3	Registration Confirmation		Mandatory
AUTH_REQ			
r_4	Authentication request for registration	Mandatory	
r_4	Authentication information for registration		Mandatory

Table 7.4 M-ESW FEAs during GMBS registration

Functional entities	FEAs	Function
Service Handler (LMT) (FE3)	932	Sends registration request to LMN.
	933	Receives authentication request from LMN following registration request.
	934	Calculates any authentication algorithm and sends authentication information to LMN.
	935	Receives confirmation of registration request.
LMN (FE4)	941	Receives registration request from LMT.
	942	Checks information in the registration request.
		Checks for user authentication. If no information is available, requests for authentication data from LMT.
	943	Receives authentication request and checks the validity of information. If authentication passed, other registration procedures are performed.
	944	Sends result of registration to LMT.

Functional Entity Actions

Table 7.4 shows the definition of the FEAs for the M-ESW Registration procedure. The table shows the different actions or functionalities executed by the M-ESW FE before and after the FE receives the signalling messages.

Similar approaches would also need to be produced for the UMTS, GPRS and W-LAN segments.

In order to complete the initial design process, the above procedures need to be repeated for each identified network functional requirement. In GMBS, this corresponds to three classes of procedures:

- GMBS generic procedures, including GMBS Registration and User Registration procedures;

- Location management, such as Location Update, Segment Selection, Segment Reselection and Session Establishment;

- Handover Management.

Many of these procedures require complex messaging sequences between a number of FEs. Once this initial phase has been completed, the next step is to consider the functional architecture (FA).

7.3   IDENTIFICATION OF FUNCTIONAL ARCHITECTURE

7.3.1   Functional Architecture

At this point, consideration is given for the first time to the spatial distribution of FEs within the network. The FA is composed of "Network Entities" (NEs) and the relations

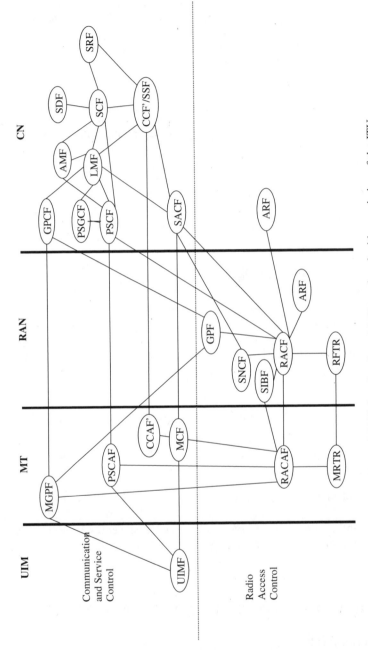

Figure 7.5 IMT-2000 functional model. Reproduced with permission of the ITU.

between them, by means of "Functional Interfaces" (FIs). An FA represents the identification of both the NEs and the FIs. A NE is obtained by grouping the FEs previously defined that belong to the same spatial area and that can be mapped onto a single item of equipment. A NE cannot be mapped onto more then one item of equipment, but conversely, more than one NE can be mapped onto one item of equipment.

The FA is used to develop the Application Layer Protocols, without considering the lower layer protocols that provide a reliable communication (by means of an underlying signalling system) between the Application Processes. These processes receive services from the Application Layer. An FA can be directly derived from the FM and this is the process that will be adopted in the following.

7.3.2 Methodology

The IMT-2000 (International Mobile Telecommunication-2000) FM [ITU-99b] is used as the reference FM to determine the GMBS FA. This is illustrated in Figure 7.5. In this section, the IMT-2000 FM will be described first before mapping the GMBS FEs onto it. The generic GMBS FA for the relevant procedures is then obtained, which is applicable to all access segments, that its UMTS, GPRS, M-ESW and W-LAN. The IMT-2000 FM will then be used to map the FEs onto the segment specific FA.

The IMT-2000 FM is divided into two parts: the Radio Access Control section and the Communication and Service Control section.

The Radio Access Control part describes the FEs related to radio link control and management. This comprises of reservation of radio resources, radio link and radio environment monitoring, handover initiation and handover execution.

The FEs belonging to the Communication and Service Control part of the IMT-2000 FM are in charge of the access control, service control, call and connection control.

The IMT-2000 FM comprises of four functional subsystems, namely UIM (User Identity Module), MT, RAN and CN (Core Network), as defined in ITU Recommendation Q.1701 [ITU-99a]. Note: MT defined in this section refers to the GMMT; RAN refers to all of the different types of RAN used in GMBS, it does not specifically refer to UMTS, a family member of the IMT-2000.

Within each functional subsystem, several FEs are identified. For example, within the MT Radio Access Control Part there are two FEs, namely RACAF and MRTR.

The names of the FEs contained within the IMT-2000 FM are listed in Table 7.5 and further information on their functionalities can be found in [ITU-99b].

The FEs within the IMT-2000 FM that are of particular relevance to the GMBS FM are UIMF, PSCAF, RACAF, MRTR and PSCF.

7.3.3 Mapping of the GMBS Functional Model on to the IMT-2000 Functional Model

The majority of the FEs defined for GMBS are mapped onto the MT, as most procedures are initiated or executed by the terminal.

Figure 7.6 illustrates the complete set of FEs identified for GMBS. It can be seen that there are no GMBS FEs mapped onto the RAN part of the model.

Table 7.5   IMT-2000 functional entities

FE	Name
AMF	Authentication Management Function
ARF	Access link Relay Function
CCAF'	Call Control Agent Function (Enhanced)
CCF'/SSF	Call Control Function (Enhanced)/Service Switching Function
GPCF	Geographic Position Control Function
GPF	Geographic Position Function
LMF	Location Management Function
MCF	Mobile Control Function
MGPF	Mobile Geographic Position Function
MRTR	Mobile Radio Transmission and Reception
PSCAF	Packet Service Control Agent Function
PSCF	Packet Service Control Function
PSGCF	Packet Service Gateway Control Function
RACAF	Radio Access Control Agent Function
RACF	Radio Access Control Function
RFTR	Radio Frequency Transmission and Reception
SACF	Service Access Control Function
SCF	Service Control Function
SDF	Service Data Function
SIBF	System access Information Broadcast Function
SNCF	Satellite Network Control Function
SRF	Specialised Resource Function
UIMF	User Identification Management Function

In Figure 7.6, FEs have been grouped together to form NEs. The lines connecting the different NEs represent the FIs between them.

Table 7.6 lists the mapping between FEs. Here, it can be seen that thirty-eight FEs are associated with the generic functional procedures. In the simple example that was used at the start of this Chapter to illustrate the development of the FEs and IFs, two of the GMBS FEs have been presented, namely RHT and CLI. From Table 7.6, it can be seen that both FEs map onto the PSCAF entity of the ITU FM.

Table 7.6   Mapping of GMBS functional entities onto IMT-2000 functional entities

IMT-2000 FE	Mapping of GMBS FE
UIMF	AIHU, HUPN, HUPU, RHT_{GSIM}, RUPU, SUPU
PSCAF	AC, AHT, CLI, CPT, EHT, HC, HCA, HD, HI, HOC, HUPN, LUH, QEF, QSP, RC, RHT_{TE}, RHT, SC, SCA, SEC, SHRU, SSH, SSSR
RACAF	TCCU
MRTR	MEF
PSCF	AHN, AIHN, EHN, LUH_N, RHN, SHRN, TCCN

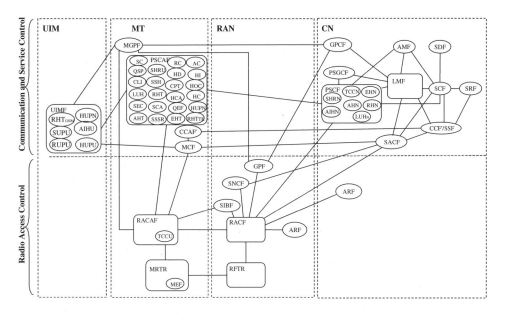

Figure 7.6   Overall mapping of functional entities onto the IMT-2000 functional model.

A brief description of the roles of each of the GMBS FEs listed in Table 7.6 follows:

Address Control (AC). This entity is in charge of managing the internal address modification following a successful Mobile IP registration. This indicates that the Care-of-Address (CoA) of the terminal would be stored locally in a database located in the Terminal Equipment (TE) itself and AC is in charge of modifying the address in the TE.

Authentication Handler Network (AHN). AHN assists AHT in performing any terminal identification functions.

Authentication Handler Terminal (AHT). The AHT performs terminal identification functions. This entity is usually located in the terminal itself.

Authentication and Integrity Handler Network (AIHN). AIHN performs user identification and integrity functions on the network.

Authentication and Integrity Handler User (AIHU). The AIHU performs user identification functions and integrity functions on the terminal.

Current Location Information. CLI is responsible for receiving information on the availability of each segment from RHT. This information is then passed to LUH to be processed.

Control Point Transfer (CPT). This entity is used to trigger Mobile IP registration by sending the latest binding updates to its home network.

Encryption Handler Network (EHN). EHN is responsible for any encryption procedures on the radio path from the network side.

Encryption Handler Terminal (EHT). EHT is responsible for any encryption procedures on the radio path from the terminal side.

Handover Criteria (HC). This entity provides HI with the criteria needed to initiate a handover. The criteria consist of several thresholds for all the parameters measured by the MEF. Thus, these criteria change on the basis of the current active segment.

Handover Criteria Adjustment (HCA). This entity sets and updates the handover criteria contained in HC, according to the instructions received from resource control functionality and network management functionality.

Handover Decision (HD). This entity decides on the target segment following a request from HI.

Handover Initiation (HI). This entity processes the information provided by the MEF entity to decide whether the criteria to initiate a handover are fulfilled. If so, HI checks the information provided by the entities HUPU and TCCU to identify the need for a handover and the target cell(s)/spot-beam(s).

If a handover is needed, HI issues an execution request to the HD entity, together with information that is needed by HD to identify the new handover control point.

HI can be located at the terminal side (if Mobile Controlled Handover (MCHO) is employed) or at both the terminal side and the network side (if Mobile Assisted Handover (MAHO) is employed; in the latter case, the HI entities at the terminal and the network side co-operate with each other).

Handover Execution Control (HOC). This entity is in charge of controlling the handover process. It co-ordinates the activities of all the entities involved in the handover execution.

Handover User Profile—Network (HUPN). This entity contains a subset of the user and service profile concerned with QoS and bandwidth requirements. The HD entity uses this information to negotiate with the user the offered QoS in the new segment.

Handover User Profile—User (HUPU). This entity contains a subset of the user profile related to the handover process. This subset contains information regarding QoS and bandwidth requirements, priority list, access rights and so on. This entity provides HI with information on the subscriber's profile so that HI can decide on whether a handover can be initiated.

Location Update Handler (LUH). LUH is responsible for maintaining track of the number of segments available. It is also responsible for passing this information to SSH and HI as segment availability changes. LUH is also responsible for storing the result of the segment selection.

Location Update Handler—Network (LUH$_N$). LUH$_N$ is the network equivalent to LUH in the T-IWU, and is responsible for storing results of the handover and segment selection.

Measurement Function (MEF). This entity measures parameters on the active link (in the active segment). These parameters relate to the quality of the signal received (in terms of bit error rate/packet error rate and/or signal strength) and to the delay affecting the radio link (in terms of overall delay, delay jitter, etc.). The type of parameters to be measured depends on the specific active link.

The measured parameters are sent to HI periodically. MEF can be located at the terminal side (if MCHO is employed) or at both the terminal side and the network side (if MAHO is employed).

Quality Evaluation Function (QEF). This entity is in charge of evaluating the overall QoS that the user perceives on the active link. This information is sent to HI periodically.

Quality of Service Provision (QSP). This entity is used to evaluate the QoS available in each active segment. QoS information is sent to SSH periodically.

Resource Control (RC). This entity requests for IP Connectivity (PDP (Packet Data Protocol) Context for GPRS) in the target segment based on the instructions received from SEC. RC is actually part of the Session Handler.

Registration Handler-Network (RHN). RHN assists RHT in registering a user/users with the T-IWU.

Registration Handler Terminal (RHT). RHT is used for handling registration procedures.

Registration Handler Terminal-GSIM (GMBS Subscriber Identity Module) (RHT_{GSIM}). RHT_{GSIM} assists RHT in performing registration procedures in the GMBS Smart Card.

Registration Handler-Terminal Equipment (RHT_{TE}). This entity is responsible for issuing a user registration request to the T-IWU when a user registers with the TE. This entity is located in the TE.

Registration User Profile—User (RUPU). This entity stores user related information such as user identification.

Segment selection Criteria (SC). SC obtains the main criteria for the segment selection.

Segment selection Criteria Adjustment (SCA). SCA sets and updates the segment selection criteria found in SC, according to the instructions received from the user. Using this entity, the user can decide on the type of service required, therefore, making it possible for SSH to initiate a segment reselection based on resource management and QoS issues.

Session Establishment Controller (SEC). SEC controls all Session Establishment procedures. It is also responsible for releasing an active IP connection and for updating the list of addresses through AC.

Special Handover Request—Network (SHRN). This entity forces HD to trigger a handover. The requests from SHRN are related to requests issued by the network due to system management issues (such as resource utilisation, maintenance, etc.) or service issues.

Special Handover Request—User (SHRU). This entity forces HI to initiate a handover. The requests coming from SHRU are related to requests issued by the user due to perceived QoS, service availability, tariff structure and so on. These issues are not directly related to radio link quality, which is evaluated by MEF.

Segment Selection Handler (SSH). This entity is used to determine the active segment based on the information provided by the other entities. It is also responsible for sending a session establishment request to SEC to trigger the GMBS Session Establishment.

Special Segment Selection Request (SSSR). This entity is used to enable a segment reselection request under special conditions such as user's requirements and issues related to network resource management. This entity can also be used to send messages to SSH when a user needs to send data but no active segments are available.

Segment Selection User Profile—User (SUPU). This entity contains a subset of the user profile related to the segment selection process. Information that can be derived from this subset includes QoS and bandwidth requirements, priority list, access rights, and so on.

Target Cell and Connection—Network (TCCN). This entity provides HD with information of resources available in the target segment(s). The HD uses this information (along with the information coming from HUPN) to negotiate the QoS to be offered to the user.

Target Cells and Connection—User (TCCU). This entity provides HI periodically with a list of target cell(s)/spot-beam(s) for each SS-MT, along with data (radio link quality, capacity) given for each element (cell/spot-beam). In the active segment, the target elements are the neighbour cell(s)/spot-beam(s). Thus, the information gathered by TCCU can be used to evaluate the QoS obtainable in the target segment(s).

7.3.4 Satellite Specific Functional Architecture

The methodology for deriving this architecture is similar to the methodology defined for deriving the generic GMBS FA. However, in this architecture, instead of using the GMBS FEs, segment specific FEs that communicate with GMBS entities are grouped together and mapped onto the IMT-2000 FM. The following FEs are introduced specifically to take into account the functionality of the satellite network.

Connection Control-Network (CCN): This entity performs connection admission control. The CCN resides in the M-ESW NOC.

Connection Control-Terminal (CCT): This entity is in charge of handling connection set-up and connection termination. It resides at the terminal side, both in the SAT-MT and in the M-ESW Fixed Earth Station (FES).

Location Management-Terminal (LMT): This is a M-ESW specific entity coinciding with the Service Handler introduced previously (see IF of Figure 7.4). It handles, by co-operating with the corresponding FE at the network side (LMN; see below) the M-ESW Registration procedure. It is located in the SAT-MT.

Location Management-Network (LMN): This is a M-ESW specific entity, located in the M-ESW NOC, which handles the M-ESW Registration procedure.

Resource Management-Payload (RMP): This pertains to the dynamic assignment and release of resources for a connection once it has been established. It resides within the Traffic Resource Manager located in the satellite payload.

Resource Management-Terminal (RMT): This pertains to the dynamic assignment and release of resources for a connection once it has been established. It resides at the terminal side both in the SAT-MT and in the M-ESW FES. This entity is activated when a data transfer on the M-ESW connection has to take place.

Spot-Beam Identifier Handler (SIH): This entity provides the TCCU with the information regarding the spot-beam where the SAT-MT is currently roaming.

The segment specific functional architecture for GPRS, UMTS and M-ESW is shown in Figure 7.7.

7.4 NETWORK ARCHITECTURE

7.4.1 Introduction

The network architecture (NA) consists of physical entities (PEs).

At this stage, systems are partitioned into several PEs by grouping together FEs that could be mapped onto a single physical location or item of equipment. The NA basically acts as a physical realisation of the system. This is also equivalent to Step 6 defined in Q.65 [ITU-00].

Relationships between PEs are defined by physical interfaces (PIs). A PI has to take into account the physical aspects of the interface and the message formats. This differs from the FI described in the previous section where only the Application Layer Protocols are considered. Although a PE is usually mapped from a NE, a PE may contain more functionality compared to a NE, such as routing.

The NA shown in Figure 7.8 represents the mapping of the IMT-2000 FM onto the M-ESW satellite architecture. This illustrates where the FEs would be located in the satellite access segment. Here, the M-ESW NOC and the FES incorporate the RAN components of the IMT-2000 FM, together with most of the CN FEs.

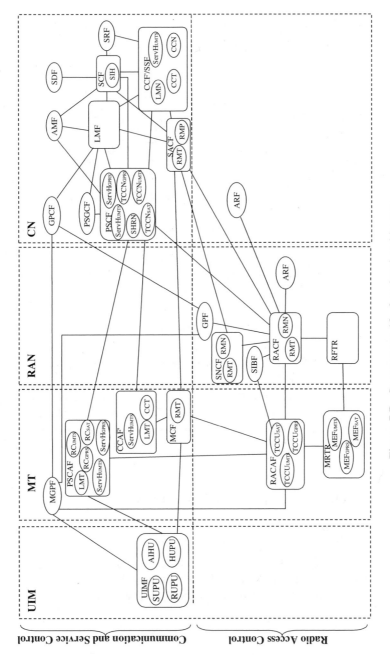

Figure 7.7 Segment specific functional architecture.

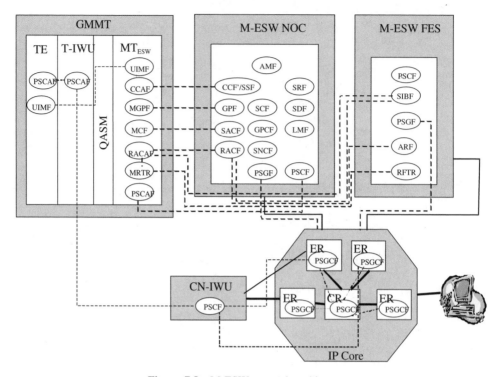

Figure 7.8 M-ESW network architecture.

It can be seen that the NA brings together the different items of equipment within the network.

The GMBS NA is shown in Figure 7.9. Here, it can be seen that the MT, previously used in the FM, has been divided into TE, T-IWU and SS-MT, corresponding to the physical items of equipment. Similarly, the GMBS Smart Card takes on the role of the User Service Identity Module. In the diagram, it can be seen that each SS-MT has its own associated MEF, RC and TCCU FEs.

7.4.2 Information Flow between Physical Entities

7.4.2.1 Signalling Protocols

Based on the PI and FI derived from the FA and NA, the signalling flow and the message format can be defined. This is analogous to Stage 3 of the ITU 3-stage approach where the messages needed to support the IFs and modifications to existing IFs between the nodes are identified [ITU-88].

7.4.2.2 GMBS Registration

Figure 7.10 shows the IF between the PEs for GMBS Registration. When the GMMT is first switched-on, the T-IWU sends a message to the SS-MTs to determine whether a segment is

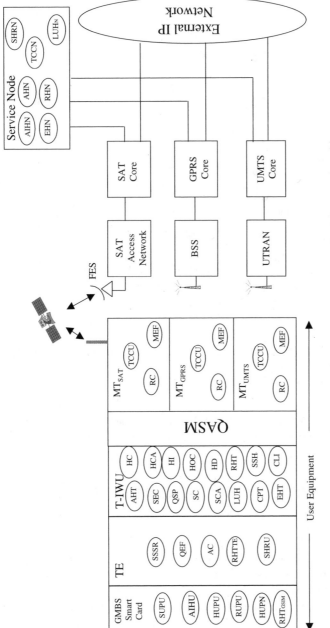

Figure 7.9 Overall mapping of inter-segment FM onto GMBS network architecture.

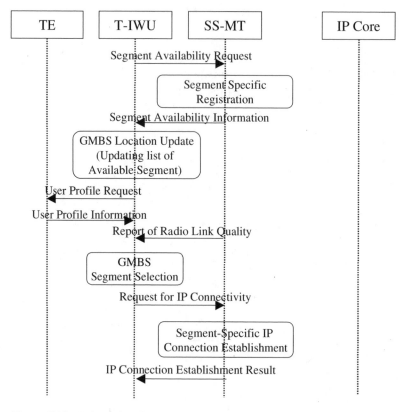

Figure 7.10   Information flow between physical entities for GMBS registration.

available or not. Following this, the GMBS Location Update and Segment Selection procedures are executed before establishing an IP connection to the core network.

7.5   FORMAL PROTOCOL SPECIFICATIONS USING SDL

7.5.1   Methodology

The main objective of this activity is to validate the signalling protocols that have been defined. This represents the final stage in the design process and is also equivalent to Step 5 of Q.65 [ITU-00], which is optional.

To achieve this, SDL is used. SDL is an object-oriented, formal language defined by the ITU under Recommendation Z.100 [ITU-99c]. The language is used for specifying complex, event-driven, real-time and interactive applications using discrete signals.

SDL consists of a set of extended finite state machines that run in parallel. It consists of four main hierarchical levels, namely System, Block, Process and Procedure. The System can be partitioned into Blocks, which are further divided into Processes and Procedures. A

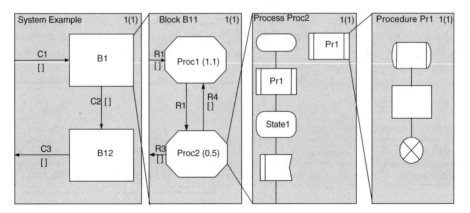

Figure 7.11   Structural overview of an SDL system.

Process, which is the basic unit of SDL, can receive and send out signals, assign values to variables, make decisions and use timers to enable real-time signalling. Figure 7.11 illustrates the structural overview of an SDL system.

The approach used in designing the GMBS SDL system is to model each PE at the block level. These blocks are then further partitioned to create FE blocks. The procedures defined in each FE are modelled at the process and procedure levels.

Five layers of hierarchy have been created for GMBS, with the PEs at the system level, followed by the IMT-2000 FEs and the GMBS specific FEs. The last two hierarchies consist of processes and procedures. The whole GMBS SDL system also communicates with the GMBS environment (e.g. user, service provider, etc.) via SDL signals. This is illustrated in Figure 7.12.

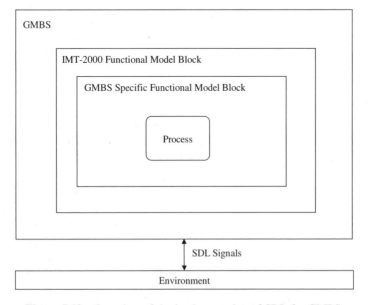

Figure 7.12   Overview of the implementation of SDL for GMBS.

In the following examples, a graphical description of SDL will be employed to illustrate the design process. This is known as SDL/GR, where GR stands for Graphical Representation. SDL makes use of a number of symbols at various stages in the design, and those that are employed in this Chapter are summarised in Table 7.7.

Table 7.7   Example symbols used in SDL

Pictorial representation	Symbol	Description
Block Name	Block reference	The name of the block is enclosed within the symbol. These symbols, connected together, and to the environment, via channels, are used to describe the system architecture.
Process Name	Process reference	The name of the process is enclosed within the symbol. These symbols are connected together, and to the environment, via channels and are used to describe the interaction between processes within blocks.
	State	The state symbol describes the state of the process model.
	Same Next state	This is used to indicate that the following next state is the same as the originating state.
	Task	A task symbol is used to assign values to data and to set and reset timers.
	Text	This symbol contains definitions of the data, signals, signal lists and comments (denoted by text within /* */).
	Decision	This symbol represents a logical decision process and allows flow to be directed down different branches based on the relevant condition.
	Connect	This symbol is used to direct flow of control to a point indicated by the connector symbol.
	Procedure start	This symbol defines the start of the procedure.
	Process start	This symbol defines the start of the process.
	Output	This symbol is used to output a named signal to a desired destination. Parameters' values can also be sent but must have values assigned to them, otherwise they will arrive undefined.
	Input	This symbol is used to identify the input signal or timer, together with a list of variable parameters.
	Procedure call	This symbol identifies the named procedure to be called, together with a list of parameters to be passed to/from the procedure.
	Exit procedure	This symbol is used to stop a procedure and return to its corresponding process.
	Exit process	This symbol is used to stop a process.

7.5.2 Message Formats

The message formats for the signals are defined using the textual version of SDL (SDL/PR), where PR stands for Phrase Representation. These are then pasted into a text symbol in SDL/GR.

AID
Source
Destination
Message Parameters
Operator

Figure 7.13 Packet format.

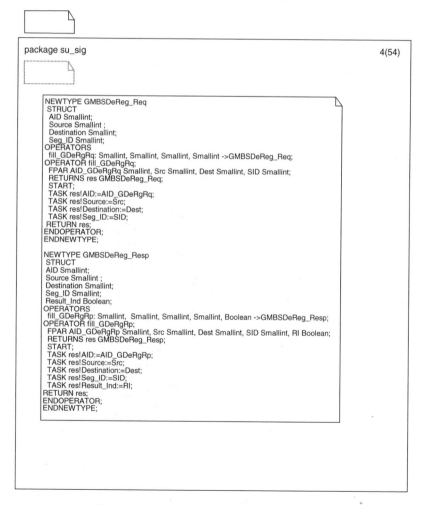

```
package su_sig                                                    4(54)


  NEWTYPE GMBSDeReg_Req
  STRUCT
   AID Smallint;
   Source Smallint ;
   Destination Smallint;
   Seg_ID Smallint;
  OPERATORS
   fill_GDeRgRq: Smallint, Smallint, Smallint, Smallint ->GMBSDeReg_Req;
  OPERATOR fill_GDeRgRq;
   FPAR AID_GDeRgRq Smallint, Src Smallint, Dest Smallint, SID Smallint;
   RETURNS res GMBSDeReg_Req;
   START;
   TASK res!AID:=AID_GDeRgRq;
   TASK res!Source:=Src;
   TASK res!Destination:=Dest;
   TASK res!Seg_ID:=SID;
   RETURN res;
  ENDOPERATOR;
  ENDNEWTYPE;

  NEWTYPE GMBSDeReg_Resp
  STRUCT
   AID Smallint;
   Source Smallint ;
   Destination Smallint;
   Seg_ID Smallint;
   Result_Ind Boolean;
  OPERATORS
   fill_GDeRgRp: Smallint, Smallint, Smallint, Smallint, Boolean ->GMBSDeReg_Resp;
  OPERATOR fill_GDeRgRp;
   FPAR AID_GDeRgRp Smallint, Src Smallint, Dest Smallint, SID Smallint, RI Boolean;
   RETURNS res GMBSDeReg_Resp;
   START;
   TASK res!AID:=AID_GDeRgRp;
   TASK res!Source:=Src;
   TASK res!Destination:=Dest;
   TASK res!Seg_ID:=SID;
   TASK res!Result_Ind:=RI;
  RETURN res;
  ENDOPERATOR;
  ENDNEWTYPE;
```

Figure 7.14 Specification of message formats.

For every definition of a signal, an Association Identity (AID), the source address, the destination address, the message parameters and an operator are inserted to simplify the process of SDL specification. The format is shown in Figure 7.13.

An example of the textual algorithmic notation of the signal is shown in Figure 7.14. In this figure, it can be seen that two different signals have been defined, namely *GMBSDe Reg_Req* and *GMBSDeReg_Resp*. Note that the package constructs have been used in the definition of message formats to introduce modularity and genericity into the models.

Messages that are directed to the same FEs are then grouped together using the ASN.1 CHOICE command. A CHOICE is similar to a Union in the C language. CHOICE allows grouping of different data structures in a single type. On the left-hand side of Figure 7.15, three different frame types are first defined for the TE, namely the GMBS_RHTTE_-FRAME, GMBS_SSSR_FRAME and GMBS_AC_FRAME. Since CHOICE is used for each frame, only one of the signals defined in the frame is present at one particular time. The frames are then used to form the main signals used for GMBS. This is shown in the box located on the right-hand side of Figure 7.15.

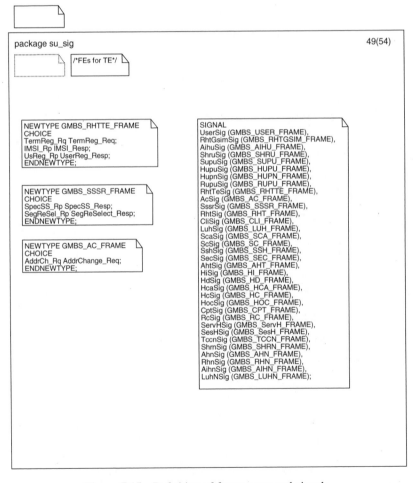

Figure 7.15 Definition of frame types and signals.

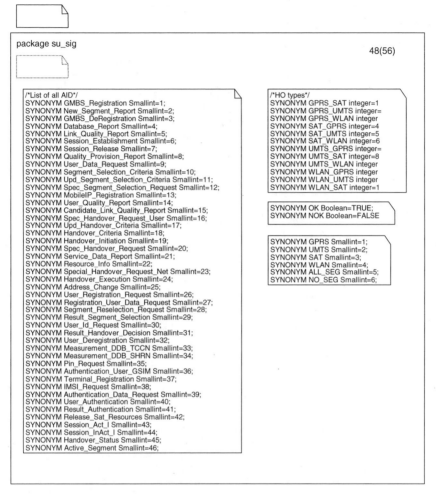

Figure 7.16 Use of synonyms in SDL.

Different synonyms and constants are also defined for GMBS. An example of this declaration can be seen in Figure 7.16, where the AIDs for each procedure are defined.

7.5.3 System Specification

The topmost level of abstraction of an SDL specification is the system level. At this level, a very abstract view of the system is obtained without going into too much detail. A system defines a set of blocks that communicate with each other and with the environment via channels.

Figure 7.17 shows the system specification for GMBS. From the diagram, it can be seen that the system consists of PEs that communicate with each other to complete all GMBS mobility procedures. Thirteen different PEs are built, corresponding to the GSIM, TE,

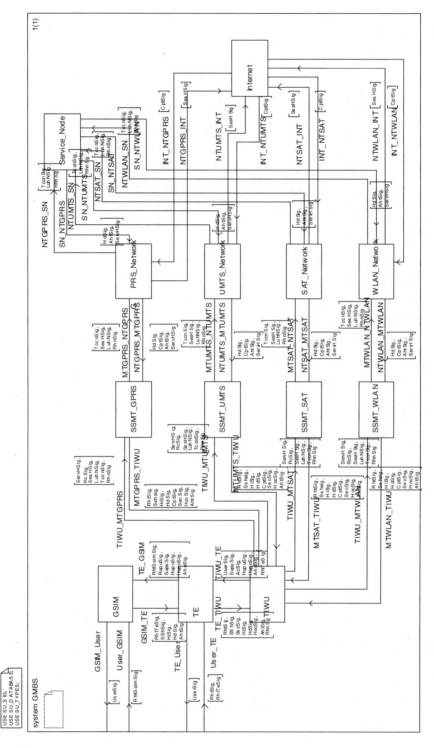

Figure 7.17 Overview of the implementation of SDL for GMBS.

Table 7.8 Example information flows between RHT$_{TE}$ physical entities

Physical entities	In	Channel	Out	Channel
T-IWU	AhtSig	TE_TIWU	AcSig	TIWU_TE
	HdSig		AihuSig	
	HiSig		HupnSig	
	HocSig		HupuSig	
	RhnSig		RhtTeSig	
	RhtSig		SupuSig	
	SecSig		UserSig	
	SshSig			
	AhtSig	MTGPRS_TIWU	LuhNSig	TIWU_MTGPRS
	CptSig	MTUMTS_TIWU	RcSig	TIWU_MTUMTS
	HdSig	MTSAT_TIWU	RhnSig	TIWU_MTSAT
	HiSig		ServHSig	
	HocSig		SesHSig	
	RhtSig		TccnSig	
	SecSig			
	SshSig			

T-IWU, SS-MTs, SS-Networks, GMBS Service Node and the Internet Network. Lines connecting the blocks indicate channels, the direction of communication being shown by an arrow. The name of the channel, e.g. TIWU_MTGPRS, and the grouping of signals are also indicated, e.g. [ServHSig].

Table 7.8 provides examples of the signalling flows between the PEs at the system level. Instead of treating each signal as a single message, the signals have been grouped together and named according to their destination address. For example, a signal named *HiSig* actually represents a group of signals directed at the FE called HI.

7.5.4   Block Specification

7.5.4.1   Overview

Block specification has been introduced as a structuring element in SDL. It describes a set of process types that interact with each other and with the channels in the enclosing system specification via signal routes [ELL-97].

In GMBS, two levels of block specification are defined, one to describe the interaction between the relevant IMT-2000 FMs and another to define the GMBS specific FMs. In the IMT-2000 model, a block substructure specification (block partitioning) is used, in which blocks defined earlier for the system specification are divided into one or more blocks. Subsequently, the block and process specifications for the GMBS FEs are defined from the IMT-2000 blocks. In this section, the IMT-2000 models will firstly be described before describing the GMBS specific FMs. Note that in the higher level (the IMT-2000 level), the blocks are specified using block types. Block types are useful in the sense that they allow the designer to create reusable components. This is particularly useful in the GMBS as the same block types can be used when designing the segment specific components.

7.5.4.2 Block Substructure Specifications for IMT-2000 FM in the T-IWU

Referring back to Figure 7.8, it can be seen that in the T-IWU, one IMT-2000 FE is applicable for GMBS. In SDL/GR, this is represented by the block, PSCAF_TIWU, which interacts with PSCAF defined in TE and SS-MT in order to execute most of the GMBS specific procedures. Since the intelligence of the network is mostly focused here, this is also one of the most complex models. As the T-IWU is physically connected to four SS-MTs, there should exist a mechanism in which one (or more) of the active SS-MTs is (are) chosen for packet transmission. This functionality is executed by a routing function in the T-IWU block, as shown in Figure 7.18. The relevant signalling protocols between the FEs are shown in Table 7.9. The router shown in the diagram is essentially used to route the packets to the correct segment. This is achieved by reading certain header/fields in the sent packets. In the PSCAF_TIWU block diagram, further FEs are defined to handle procedures allocated to the PSCAF, such as RHT, CLI and SSH. An example of the information flow for PSCAF in the T-IWU is listed in Table 7.10.

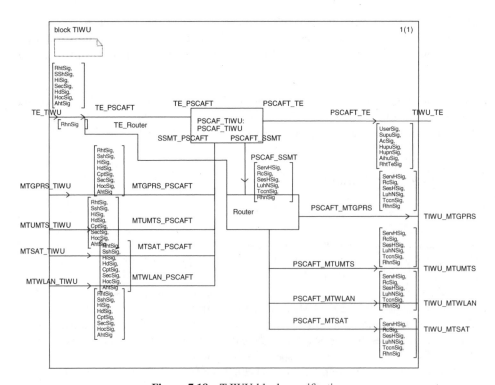

Figure 7.18   T-IWU block specification.

Table 7.9 Information flows for T-IWU

FE	In	Channel name	Out	Channel name
PSCAF_TIWU	AhtSig	TE_PSCAFT	Acsig	PSCAFT_TE
	HdSig		AihuSig	
	HiSig		HupnSig	
	HocSig		HupuSig	
	RhtSig		RhtTeSig	
	SecSig		SupuSig	
	SshSig		UserSig	
	AhtSig	MTGPRS_PSCAFT	LuhNSig	PSCAFT_SSMT
	CptSig	MTUMTS_PSCAFT	RcSig	
	HdSig	MTSAT_PSCAFT	RhnSig	
	HiSig	MTWLAN_PSCAFT	ServHSig	
	HocSig		SesHSig	
	RhtSig		TccnSig	
	SecSig			
	SshSig			
Router	LuhNSig	PSCAFT_SSMT	LuhNSig	PSCAFT_MTGPRS
	RcSig		RcSig	PSCAFT_MTUMTS
	RhnSig		RhnSig	PSCAFT_MTSAT
	ServHSig		ServHSig	PSCAFT_MTWLAN
	SesHSig		SesHSig	
	TccnSig		TccnSig	
	RhnSig	TE_Router		

7.5.5 Process Specification

7.5.5.1 Overview

The process specification deals with the dynamic behaviour of processes in a system. At this level, behaviour is specified directly for a process type without having to use further structuring concepts. Once a block diagram representation for each PE in GMBS is complete, the procedures defined for the GMBS FEs can then be built using these process diagrams. Each diagram represents the procedures executed in each FE. This section aims to show an example process specification for the FEs for the Registration procedure considered at the beginning of the Chapter. These diagrams are produced based on the specifications defined previously.

A process contains state machines and has one or more instances. The process instances run in parallel, are independent and contain four implicit variables, defined as follows:

Self: contains Process ID (Pid) of the current instance,

Sender: contains the Pid of the instance which sent the last signal input,

Parent: contains Pid of the instance which created the current instance,

Offspring: contains Pid of the instance created.

In the following, the process specification for the RHT entity will be shown.

Table 7.10 Example information flows for PSCAF in the T-IWU.

FE	In	Signal route	Gates	Out	Signal route	Gates
RHT	RhtSig	TE_RHT	TE_PSCAFT	RhtTeSig, SupuSig, UserSig	RHT_TE	PSCAFT_TE
	RhtSig	SSMT_RHT	SSMT_PSCAFT	RhnSig, ServHSig	RHT_SSMT	PSCAF_SSMT
	RhtSig	CLI_RHT		CliSig	RHT_CLI	
	RhtSig	AHT_RHT				
CLI	CliSig	RHT_CLI		RhtSig	CLI_RHT	
				LuhSig	CLI_LUH	
SSH	SshSig	SSMT_SSH	SSMT_PSCAFT			
	SshSig	TE_SSH	TE_PSCAFT	SupuSig	SSH_TE	PSCAFT_TE
	SshSig	LUH_SSH		LuhSig	SSH_LUH	
	SshSig	SC_SSH		ScSig	SSH_SC	
	SshSig	QSP_SSH				
	SshSig	SEC_SSH		SecSig	SSH_SEC	
	SshSig	RHT_SSH		RhtSig	SSH_RHT	
	SshSig	CPT_SSH				
				HiSig	SSH_HI	

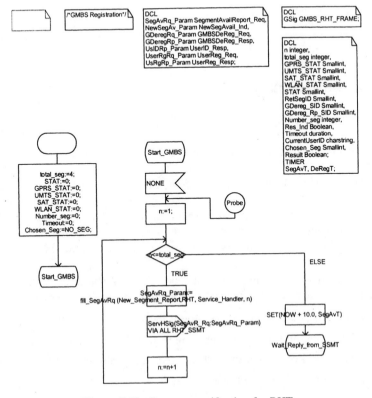

Figure 7.19 Process specification for RHT.

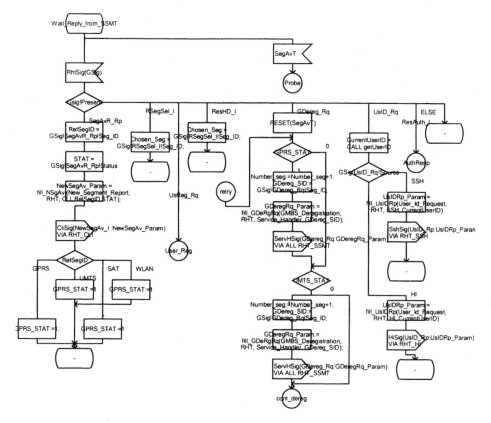

Figure 7.19   (*Continued*)

7.5.5.2   *Registration and Authentication*

An example of the process diagram obtained from the RHT FE is shown in Figure 7.19. This procedure starts executing when the GMMT is first switched-on. Immediately following the *start* symbol, the values of several variables defined at the top of the diagram are initialised before RHT goes to state *Start_GMBS*. Once it is in *Start_GMBS* state, RHT should, according to the specifications, probe the segment specific terminals for segment availability. However, in Figure 7.19, the transition following the state is a spontaneous transition. This is because in SDL, the only legal symbol after the state is the input, save or continuous signal. If an action is to be performed without consuming any signal, the only solution is to use a spontaneous transition, which is actually an input signal containing NONE. A loop is then used to send the *SegAvR_Rq* message to probe the availability of segments. Once this is complete, a timer, known as the *SegAvT* timer is set to interrupt RHT after 10 seconds. Following this, RHT goes to state *Wait_Reply_ - from_SSMT*.

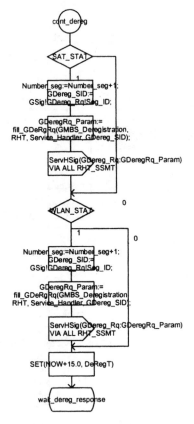

Figure 7.19   (*Continued*)

In this new state, two main inputs are possible. One is the *SegAvT* input, which will cause the RHT to probe each segment again in order to update the status of each segment. Other than that, RHT may also receive a signal known as *RhtSig*. This signal is actually capable of carrying several types of message. However, since the command *CHOICE* has been used, only one of these messages can be active at one particular time. *Gsig* in the diagram is defined to be of type GMBS_RHT_FRAME. This variable is then checked to determine the type of messages received by RHT. In this state, RHT is capable of receiving seven different types of message, namely *SegAv_Rp*, *RsegSel_I*, *UsReg_Rq*, *ResHD_I*, *Gdereg_Rq*, *UsID_Rq* and *ResAuth_I*. All other messages will be discarded.

Once the messages have been determined, RHT then proceeds to perform any other task required. In RHT, when a *SegAvR_Rp* message is obtained, RHT updates the status of the segments. Then it proceeds to send the status of the segment to CLI to be processed. When RHT receives either *RSegSel_I* or *ResHD_I*, it updates the database of the selected segment. *UsReg_Rq* and *ResAuth_I* are essentially used for user registration and authentication procedures, whereas *UsID_Rq* is used for obtaining the current user ID. However, when

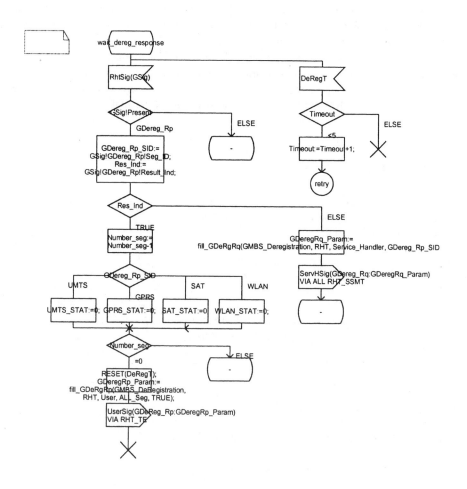

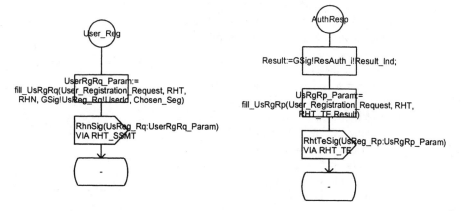

Figure 7.19 (*Continued*)

RHT receives a *GDereg_Rq* message, it first checks the status of the access segments before sending the *GDereg_Rq* message to the appropriate segment specific entities. It then goes into state *wait_dereg_response* before terminating any RHT procedures.

Upon completion of the design, the next phase is to implement the real-time software associated with the protocols and to perform prototype development to allow performance to be evaluated in realistic operating conditions. Many of the commercially available SDL design tools automatically generate real-time software from the SDL representation.

REFERENCES

[ELL-97] J. Ellsberger, D. Hoegrefe, A. Sarma: SDL: Formal Object-Oriented Language for Communicating Systems, *Prentice Hall*, Great Britain, 1997.

[ITU-00] ITU-T Recommendation Q.65, The Unified Functional Methodology for the Characterization of Services and Network Capabilities, (06/2000).

[ITU-88] ITU-T Recommendation I.130, Method for the Characterization of Telecommunication Services Supported by an ISDN and Network Capabilities of an ISDN, (11/1988).

[ITU-99a] ITU-T Recommendation Q.1701, Framework for IMT-2000 Networks, (03/1999).

[ITU-99b] ITU-T Recommendation Q.1711, Network Functional Model for IMT-2000, (03/1999).

[ITU-99c] ITU-T Recommendation Z.100, Specification and Description Language, (11/1999).

[ITU-99d] ITU-T Recommendation Z.120, Message Sequence Chart, (11/1999).

ACRONYMS

AAC	Authentication and Access Control
AC	Address Control
AHN	Authentication Handler Network
AHT	Authentication Handler Terminal
AID	Association Identity
AIHN	Authentication and Integrity Handler Network
AIHU	Authentication and Integrity Handler User
AMF	Authentication Management Function
ARF	Access link Relay Function
B-ISDN	Broadband ISDN
CCAF'	Call Control Agent Function (Enhanced)
CCF'	Call Control Function (Enhanced)
CCN	Connection Control-Network
CCT	Connection Control-Terminal
CLI	Current Location Information
CN	Core Network
CoA	Care-of-Address
CPT	Control Point Transfer
EHN	Encryption Handler Network

EHT	Encryption Handler Terminal
ER	Edge Router
FA	Functional Architecture
FE	Function Entity
FEA	FE Action
FES	Fixed Earth Station
FI	Functional Interface
FM	Functional Model
GMBS	Global Mobile Broadband System
GMMT	GMBS Multi-Mode Terminal
GPCF	Geographic Position Control Function
GPF	Geographic Position Function
GPRS	General Packet Radio Service
GSIM	GMBS Subscriber Identity Module
HA	Home Agent
HC	Handover Criteria
HCA	HC Adjustment
HD	Handover Decision
HI	Handover Initiation
HOC	Handover execution Control
HUPN	Handover User Profile—Network
HUPU	Handover User Profile—User
IF	Information Flow
IMT-2000	International Mobile Telecommunication—2000
IP	Internet Protocol
ISDN	Integrated Services Digital Network
ITU	International Telecommunication Union
IWU	Inter-Working Unit
LE	Logical Entity
LM	Logical Model
LMF	Location Management Function
LMN	Location Management-Network
LMT	Location Management-Terminal
LUH	Location Update Handler
LUH_N	LUH Network equivalent
MAHO	Mobile Assisted Handover
MCF	Mobile Control Function
MCHO	Mobile Controlled Handover
MEF	Measurement Function
M-ESW	Mobile-EuroSkyWay
MGPF	Mobile Geographic Position Function
MRTR	Mobile Radio Transmission and Reception
MT	Mobile Terminal
NA	Network Architecture
NE	Network Entity
NOC	Network Operation Centre

PDP	Packet Data Protocol
PE	Physical Entity
PI	Physical Interface
Pid	Process Identification
PIN	Personal Identification Number
PSCAF	Packet Service Control Agent Function
PSCF	Packet Service Control Function
PSGCF	Packet Service Gateway Control Function
QASM	QoS Support Module
QEF	Quality Evaluation Function
QoS	Quality of Service
QSP	Quality of Service Provision
RACAF	Radio Access Control Agent Function
RACF	Radio Access Control Function
RAN	Radio Access Network
RC	Resource Control
REC	Reference Configuration
RFTR	Radio Frequency Transmission and Reception
RHN	Registration Handler Network
RHT	Registration Handler Terminal
RHT_{GSIM}	RHT in GSIM smart card
RHT_{TE}	RHT in Terminal Equipment
RMP	Resource Management-Payload
RMT	Resource Management-Terminal
RUPU	Registration User Profile User
SACF	Service Access Control Function
SAT	Satellite
SC	Segment selection Criteria
SCA	SC Adjustment
SCF	Service Control Function
SDL	Specification and Description Language
SDL/GR	SDL/Graphical Representation
SDL/PR	SDL/Phase Representation
SEC	Session Establishment Controller
SHRN	Special Handover Request—Network
SHRU	Special Handover Request—User
SIB	Service Independent Building blocks
SIBF	System access Information Broadcast Function
SIH	Spot-beam Identifier Handler
SNCF	Satellite Network Control Function
SRF	Specialised Resource Function
SS-MT	Segment Specific MT
SSH	Segment Selection Handler
SSSR	Special Segment Selection Request
SUPU	Segment Selection User Profile—User
T-IWU	Terminal IWU

TCCN	Target Cell and Connection Network
TCCU	Target Cell and Connection User
TE	Terminal Equipment
UFM	Unified Functional Methodology
UIM	User Identity Module
UIMF	User Identification Management Function
UMTS	Universal Mobile Telecommunications System
W-LAN	Wireless Local Area Network

8

Performance Validation

NICOLA BLEFARI-MELAZZI[1], FABRIZIO CEPRANI[2],
MATTHIAS HOLZBOCK[3], AXEL JAHN[3],
VINCENZO SCHENA[2] and FABIO VECCHIA[2]

*[1]University of Rome "Tor Vergata", Italy; [2]Alenia Spazio, Italy; [3]German Aerospace Center,
Germany*

8.1   INTRODUCTION

The study phase, as outlined in the previous Chapters, has resulted in the overall design of the Global Mobile Broadband System (GMBS) reference network architecture and its capabilities. In this concluding Chapter, details are provided of the demonstration activity, the purpose of which is to validate the designed protocols and to evaluate their performance in realistic transmission environments.

The demonstration phase is characterised by four main activities:

1. Development of the validation strategy and definition of test scenarios;

2. Design and development of a suitable demonstrator;

3. Execution of the demonstrator campaigns;

4. Performance evaluation through experimentation.

The GMBS demonstrator, in line with the theoretical analysis, is based on three wireless segments:

- Mobile EuroSkyWay (M-ESW) satellite segment,

- GPRS (General Packet Radio Service) and UMTS (Universal Mobile Telecommunications System) terrestrial segments,

- Wirelesses Local Area Network (W-LAN) segment.

Before the demonstration phase can be performed, the first step is to develop the validation strategy.

Space/Terrestrial Mobile Networks.   Edited by R.E. Sheriff, Y.F. Hu, G. Losquadro, P. Conforto, C. Tocci
© 2004 John Wiley & Sons, Ltd   ISBN: 0-470-85031-0

8.2 VALIDATION STRATEGY

8.2.1 Methodology

The validation strategy adopts a fourfold experimental approach:

1. The communication capabilities of the GMBS demonstrator are validated and evaluated, by composing a subset of those envisaged in the GMBS reference network:

   End-to-End (E2E) Internet Protocol (IP) Networking. This aims to evaluate the capability of the GMBS platform to provide Internet access and IP connectivity in wireless environments, exploiting both terrestrial and satellite radio coverage. Mobile IP signalling protocols are taken into account, thus also validating the Control Plane.

   Inter-Segment IP Mobility Support. This aims to evaluate the performance of the GMBS platform for what concerns its capacity to provide both the access and the information transfer to the Internet. This demonstration mainly deals with the User Plane.

   E2E Quality of Service (QoS) Guarantee. Both User and Control Planes are addressed and validated.

2. A *step-by-step methodology* is adopted to demonstrate the selected communication capabilities.

3. The selection of a representative set of both *IP packet flows* and *IP-based applications* (differing in their traffic profile, user requirements and the related network performance constraints).

4. The identification of *performance parameters* that need to be measured to assess the GMBS capability to support IP networking, multi-segment IP mobility and E2E QoS.

8.2.2 Performance Metrics

To quantify the network performance during the execution phase and to be compliant with the validation strategy outcomes, Internet packet routing requires the characterisation of the following major performance parameters:

- Throughput: the maximum data transmission rate that can be sustained for a particular link;

- Reliability: a measure of transmission errors and packet loss measured within a specified time interval;

- Delay: the time taken by a packet to travel from source to destination;

- Jitter: the variation in E2E delay.

Taking into account the above and referring to the ITU-T Recommendation I.380 [ITU-99], the crucial performance parameters considered during the measurement and evaluation phases are:

IP packet throughput: This is the number of successful IP packets observed at the destination side within a specified time interval, divided by the time interval duration.

IP packet transfer delay (IPTD): This is the measured delay between the source and the destination point in an IP network connection for IP packets received successfully or in error.

Mean IPTD: This is the arithmetic average value of the IP packet delay calculated for an IP packet population.

IP packet delay variation (IPDV): This is a measure of the IP packet delay variation.

IP packet error ratio: This is the ratio of the total number of IP packets in error to the total number of IP packets exchanged between source and destination.

IP packet loss ratio: This is the ratio of the total number of lost IP packets to the total number of IP packets exchanged between source and destination.

Spurious IP packet rate: This is the number of spurious IP packets observed at the destination side in a particular time interval, divided by the time interval duration.

8.3   EXPERIMENTAL STRATEGY

As far as the experimental phase is concerned, two measurement campaigns are employed: the *Test-bench* and the *Overall GMBS demonstrator*.

The Test-bench measurement campaign is used to perform preliminary laboratory tests on the multi-segment network; validating firstly the overall network connectivity and, subsequently, testing measurement equipment and procedures. After the analysis of the Test-bench measurement campaign, the demonstrator equipment is configured in a realistic operational environment, including actual radio link behaviour, to validate the performance of the communication capabilities.

Figure 8.1 summarises the above approach.

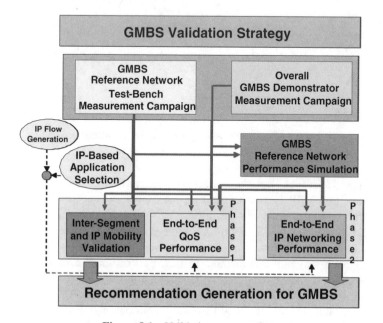

Figure 8.1   Validation strategy flow.

In the following sections, performance evaluation is addressed from the perspective of QoS.

The first validation campaign is executed in two phases: the first phase comprises of the entire E2E prototype in the laboratory. In this context, a hard-wired connection between the GMBS Multi-Mode Terminal (GMMT) and the access network substitutes for the wireless access. In the second phase, the satellite acts as the wireless access segment. The comparison between the first set of tests and the second can be used to distinguish between the delay and the packet loss caused by the satellite and that caused by the terrestrial network.

The second validation campaign is carried out in the field, having the entire set of wireless segments available and a moving mobile node installed in a van.

8.4   FIRST VALIDATION CAMPAIGN

8.4.1   First Phase

8.4.1.1   Test Configuration

During the first phase, the major group of measurements are performed on the fixed access network. Figure 8.2 depicts the probing and traffic generation set-up.

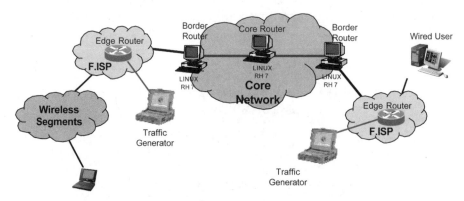

Figure 8.2   Traffic loader and analyser configuration.

The network performance tests concentrate on constant bit rate traffic. A particular audio application, based on VoIP, is used to simulate the user traffic constituting the background loading of the network. VoIP has stringent QoS conditions, specifically a maximum acceptable E2E time delay of 150 ms to guarantee good quality conversation.

The load traffic is generated randomly in order to simulate typical Internet traffic. Shaping is achieved by using an empirical method, in fact, by "sniffing" actual Internet accesses. After this operation, statistical handling of the recorded traffic is made available for loading onto the experimental network.

In the experimental network, the total bandwidth considered at the core was chosen to be 10 Mbps (typical 10Base-T LAN configuration). Taking into account hardware performance,

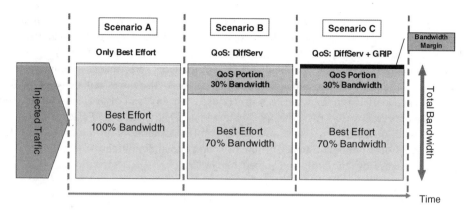

Figure 8.3 Traffic shaping.

the final available total bandwidth was about 6 Mbps. Figure 8.3 shows the bandwidth configuration for the three scenarios: Best Effort (BE), BE + DiffServ and BE + DiffServ & GRIP.

In Scenario A, all of the available bandwidth is used for BE traffic, loading the "channel" with the previously described Internet traffic.

In Scenario B, the QoS configuration with DiffServ is activated[1], reserving 30% of the total bandwidth (1.8 Mbps), while the remaining 70% is made available for BE traffic. In the reserved channel, all the QoS connections (marked by the Type of Service field in the IP header) are accepted - eventually saturating the available bandwidth. In this case, by continuing to accept connections, there is the risk of overloading the reserved channel, decreasing drastically the performance of the QoS traffic (the risk, here, is to have performance that is worse than that of the BE channel).

In Scenario C, the GRIP procedure is activated. In this way, there is a Connection Admission Control (CAC) before accepting QoS connections, thus avoiding bandwidth saturation and guaranteeing suitable performances for the accepted traffic. In this scenario, two forms of access management are possible: optimised and performed.

In the optimised bandwidth configuration, the vast majority of QoS bandwidth is set aside (e.g. about 90%) for the traffic, while the remainder is used to manager traffic bursts. In the performed bandwidth configuration, the amount of bandwidth allocated to the traffic is decreased, reducing the number of accepted connections (in this way only the high priority traffic is accepted), but by doing so, guaranteeing a very high performance to such traffic. In the two configurations, the connections that are not accepted are redirected to the BE channel (they are not rejected).

As shown in Figure 8.2, two traffic generators are used, one as source and the other as destination, each being directly connected to the ERs in the access network. These are responsible for generating background IP-based traffic, emulating a high number of simultaneous users and loading the network so as to reach saturation. Three computers are used to implement the DiffServ solution enhanced by the GRIP mechanism in the core network (the two BRs and the CR) [BIA-02]. These computers are all LINUX Operating

---

[1]RSVP resides on the network edges, starting the QoS procedures and making available the messages to manage the QoS procedures in the core network.

System (OS) based. There are two IntServ areas surrounding the core network: the first one from the Mobile User to the BR and the second from the Wired User to the other BR. The BRs implement the mapping between the different areas (IntServ and DiffServ).

In the following paragraphs, the results start with the QoS performance evaluation.

The first phase validation campaign is used to evaluate the terrestrial network performance, independently from the wireless segment used; for this kind of test, the Mobile User is directly connected to the ER. The QoS parameters (Delay, Jitter and Packet Loss) are measured with different configurations of the traffic generators in order to test the performance of the application under different network load conditions.

The performance evaluation is considered for three different test scenarios:

- *No QoS support*: In this case, only BE Internet traffic is available.

- *QoS support at the edge sub-network*: This employs an IntServ model at the edge of the sub-network and DiffServ in the core network.

- *QoS support at E2E level*: This employs an IntServ model at the edge of the sub-network and enhanced DiffServ, via GRIP, in the core network.

8.4.1.2   Results

Internet Round Trip Time (RTT)

Traffic generators were configured to create a large amount of BE traffic to use all the available bandwidth. In this way, the BE classes always have a large number of packets and so the queuing delays are quite high (as is packet loss—about 10%).

An example of typical packet delays (Round Trip Time—RTT) in the Internet network using the BE solution is reported in Figure 8.4. Delays are in the region of 100 ms (the average is 107 ms) with some very high peaks, in some cases representing delays of in excess of 1.5 s and in one particular instance, more than 2 s.

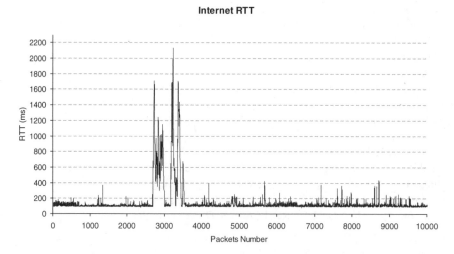

Figure 8.4   Example of typical internet packet delays.

The traffic generators and the routers in the DiffServ area were configured to have delays similar to the Internet delays, using the BE solution in the GMBS demonstrator.

In this way the following tests of the DiffServ and the GRIP solutions show the improvements that these can obtain not only for the GMBS network but also for the Internet network.

Scenario A: Fixed Access and Core Network – Best Effort (Laboratory)

Figure 8.5 shows the RTT for the application used.

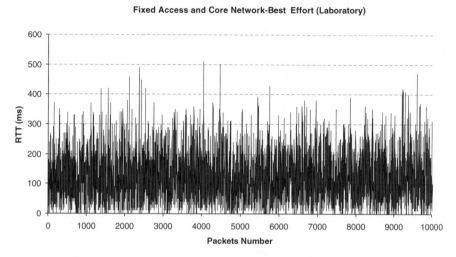

Figure 8.5 RTT—Best effort for first validation campaign first phase.

This graph is similar to the previous, even if it has not the same peaks. This is due to the fact that in the Internet network there are unpredictable situations and it is impossible to create these in the laboratory (the transmission of the traffic generators is always regular).

The average delay is 117 ms (quite near to the previous average delay—107 ms).

The following figures (Figures 8.6 and 8.7) show the E2E delay or IPTD and the IPDT or jitter.

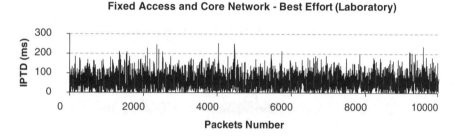

Figure 8.6 E2E Delay (IPTD)—Best effort for first validation campaign first phase.

Fixed Access and Core Network - Best Effort (Laboratory)

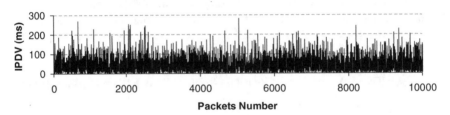

Figure 8.7   IPDV (Jitter)—Best effort for first validation campaign first phase.

Scenario B: DiffServ

To test the DiffServ solution, the traffic generators were configured to generate the previous amount of traffic, divided between the BE and the AF4 class [IET-99] (Note: the other classes could be used instead of AF4).

As expected, when there is no congestion the DiffServ solution is always better than the BE and produces very low delays.

The most interesting test for the DiffServ solution has shown that when congestion arises, the AF4 performance degrades and becomes very similar to the BE performance.

The following figures (8.8, 8.9 and 8.10) show the results of these tests when congestion arises:

Fixed Access and Core Network - DiffServ (Laboratory)

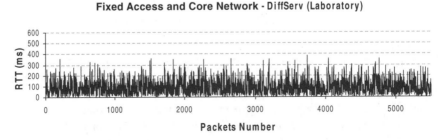

Figure 8.8   RTT—DiffServ for first validation campaign first phase.

Fixed Access and Core Network - DiffServ (Laboratory)

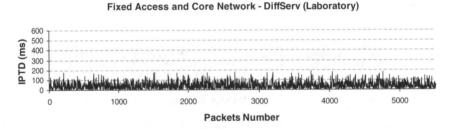

Figure 8.9   E2E Delay (IPTD)—DiffServ for first validation campaign first phase.

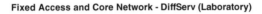

Fixed Access and Core Network - DiffServ (Laboratory)

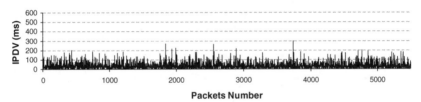

Figure 8.10 IPDV (Jitter)—DiffServ for first validation campaign first phase.

Scenario C: GRIP

This set of tests has shown that the CAC implemented by the GRIP solution provides good performance for the DiffServ classes, even when congestion arises.

The total amount of traffic generated is the same as in the previous cases.

When the number of connections accepted for each class by the GRIP mechanism is low then this solution produces very low delays but also a very low network utilisation.

The main problem is to find the number of connections to accept, for each class (in particular for the AF4 class), which produces low delays and jitter but also a high network utilisation. A good value allows a network utilisation more or less equal to 70%.

The following figures (Figures 8.11, 8.12 and 8.13) show the results of these tests:

Fixed Access and Core Network - DiffServ+Grip (Laboratory)

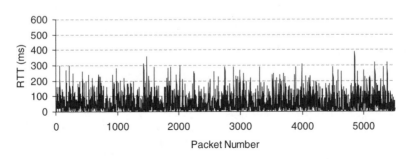

Figure 8.11 RTT—DiffServ & GRIP for first validation campaign first phase.

Fixed Access and Core Network - DiffServ+Grip (Laboratory)

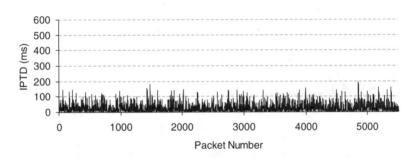

Figure 8.12 E2E Delay (IPTD)—DiffServ & GRIP for first validation campaign first phase.

Table 8.1 RTT Statistics for first validation campaign first phase

	Best effort	DiffServ	DiffServ & GRIP
Maximum value (ms)	510	501	390
Average value (ms)	117.3	88.3	56.4
Standard deviation (ms)	77.6	67.3	56.5

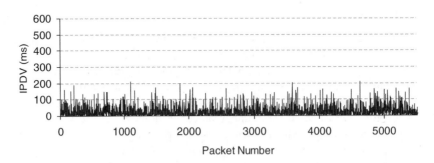

Figure 8.13 IPDV (Jitter)—DiffServ & GRIP for first validation campaign first phase.

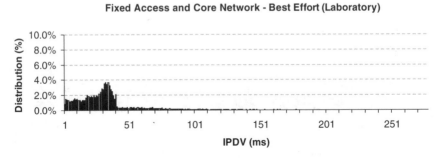

Figure 8.14 IPDV (Jitter) Distribution—Best effort for first validation campaign first phase.

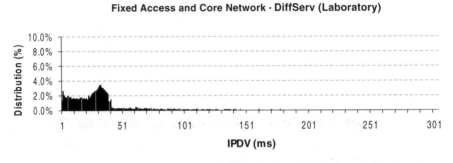

Figure 8.15 IPDV (Jitter) Distribution—DiffServ for first validation campaign first phase.

Fixed Access and Core Network - DiffServ+Grip (Laboratory)

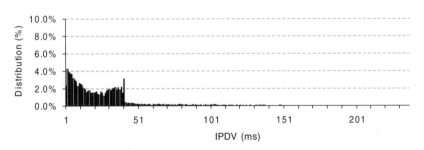

Figure 8.16   IPDV (Jitter) Distribution—DiffServ & GRIP for first validation campaign first phase.

Some significant data to compare the different solutions are reported in Table 8.1.

Figures 8.14, 8.15 and 8.16 show the Jitter distributions.

The comparison of the three previous graphs show that the DiffServ solution enhanced by the GRIP mechanism performs best. The comparison between the first and the second graph is more difficult, but it is possible to understand that the DiffServ solution performs better than the BE from Table 8.2.

Another important comparative measure is the number of packets lost in the three solutions, where the number of packets sent is always 8000. This is shown in Table 8.3.

Table 8.2   IPDV (Jitter) statistics for first validation campaign first phase

	Best effort	DiffServ	DiffServ & GRIP
Minimum value (ms)	0	0	0
Maximum value (ms)	282	304	236
Average value (ms)	35.5	30.9	24.5
Standard deviation (ms)	30.9	28.4	25.2

Table 8.3   Packet loss for first validation campaign first phase

Best effort	DiffServ	DiffServ & GRIP
Packets received: 7102	Packets received: 7495	Packets received: 7734
Packets lost: 898	Packets lost: 505	Packets lost: 266
Loss (percentage): 11.2%	Loss (percentage): 6.3%	Loss (percentage): 3.3%

8.4.2   Second Phase

8.4.2.1   Test Configuration

In the first validation campaign second phase, this time using the satellite access segment, the BE solution is compared directly with the DiffServ solution improved by the GRIP

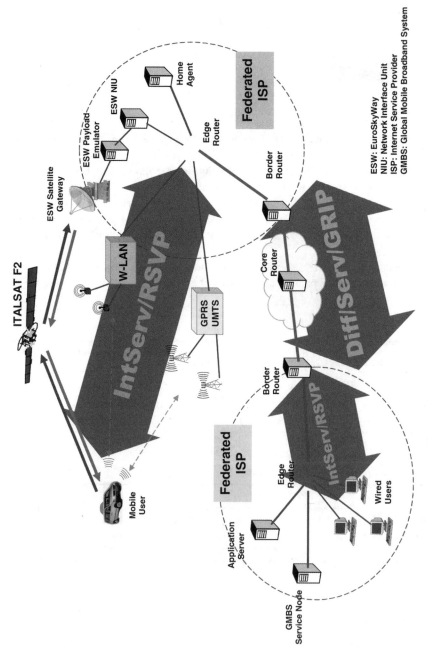

Figure 8.17 GMBS demonstrator configuration.

mechanism (the DiffServ & GRIP solution will be indicated as QoS solution in the following graphs). The pure DiffServ solution is not analysed because, in this way, it is possible to see more clearly the differences between the two solutions. The second phase demonstrator configuration is shown in Figure 8.17.

8.4.2.2   Results

Round Trip Time

The configurations of the traffic generators and the DiffServ routers are exactly the same as those of the first phase campaign. Figures 8.18 and 8.19 represent the graphs of the RTT.

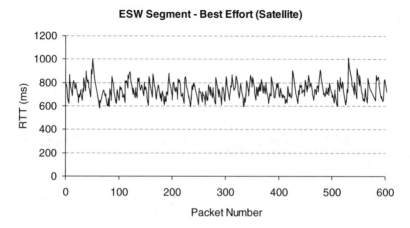

Figure 8.18   RTT—Best Effort for first validation campaign second phase.

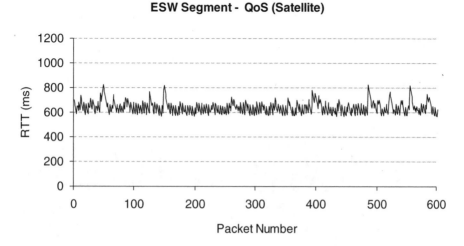

Figure 8.19   RTT—DiffServ & GRIP for first validation campaign second phase.

The RTT graphs show that by using the DiffServ solution enhanced with GRIP, the delays decrease and are roughly constant. It can be observed that the satellite delay is a little less than 600 ms, which is to be expected for a geostationary satellite.

IP Packet Delay Variation (Jitter)

The jitter graphs are presented in Figures 8.20 and 8.21.

Clearly in this case, using the DiffServ & GRIP solution, the jitter is lower because, as noted previously, the delays are approximately constant and so the delay variation is very low.

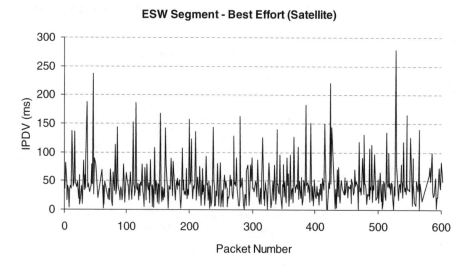

Figure 8.20   IPDV (Jitter)—Best effort for first validation campaign second phase.

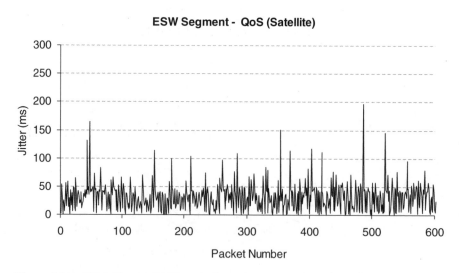

Figure 8.21   IPDV (Jitter)—DiffServ & GRIP for first validation campaign second phase.

Table 8.4 RTT statistics for first validation campaign second phase

	Best effort	QoS (DiffServ & GRIP)
Minimum value (ms)	583	567
Maximum value (ms)	1016	825
Average value (ms)	735	641
Standard deviation (ms)	72.4	47.2

Some significant data to enable a comparison between the two solutions are reported in Table 8.4.

Jitter distributions are shown in Figures 8.22 and 8.23.

By comparing the two graphs it is possible to see, in the second, a peak on the left-hand side (corresponding to delay variations very close to zero); this means that in the QoS solution delays are more consistent than those in the BE solution.

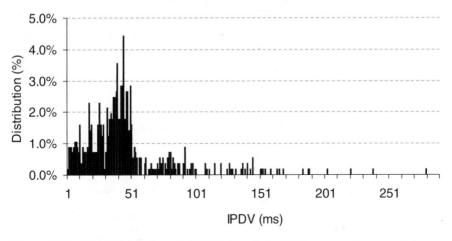

Figure 8.22 IPDV Distribution—Best Effort for first validation campaign second phase.

For the purpose of comparison, Table 8.5 is shown for this case.

Finally the number of packets lost in the two solutions is presented in Table 8.6, where the number of packets sent is 8000 in both of the solutions.

8.4.3 Performance Evaluation Conclusions

The first validation campaigns have shown that it is necessary to implement CAC in DiffServ networks because, when congestion arises, performance is very low not only in the BE class

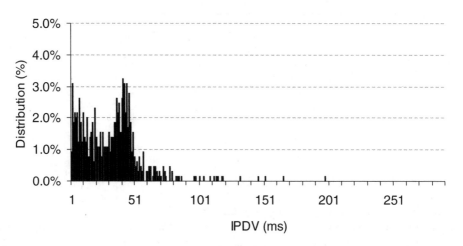

Figure 8.23 IPDV Distribution—DiffServ & GRIP for first validation campaign second phase.

Table 8.5 IPDV (Jitter) statistics for first validation campaign second phase

	Best effort	QoS (DiffServ & GRIP)
Minimum value (ms)	0	0
Maximum value (ms)	279	197
Average value (ms)	45.4	32
Standard deviation (ms)	35.9	24

Table 8.6 Packet loss for first validation campaign second phase

Best effort	QoS (DiffServ + GRIP)
Packets received: 564	Packets received: 639
Packets lost: 76	Packets lost: 1
Loss (percentage): 11.8%	Loss (percentage): 0.1%

but also in the DiffServ classes. The implementation of the GRIP mechanism shows that it provides good performance, even when congestion arises.

During the first phase campaign, the GRIP, DiffServ and BE solutions were tested, independently from the access segment used, while in the second phase the same solutions were tested using the satellite access segment.

Results have shown that:

- The BE solution produces very high loss and delays when the network is congested, so it is not a good solution, especially for real-time applications.

- The DiffServ solution is better than the BE when the DiffServ classes are not congested, but the performance becomes very low when congestion arises in the DiffServ classes (independently from the class considered).

- The GRIP Admission Control provides good performance even when the network is congested and can be configured to satisfy the requirements of real-time applications.

8.5 SECOND VALIDATION CAMPAIGN

8.5.1 Test Configuration

8.5.1.1 Constituents

Two main physical sections constitute the second phase validation demonstrator:

- The mobile section called GMMT. The GMMT constitutes the mobile component in the demonstrator layout;
- The fixed section, called Fixed Network Segment. The fixed part of the GMBS demonstrator layout provides the GMMT with access to the Internet infrastructure, guaranteeing a defined quality in the access (QoS) and furnishing all the mobility management capabilities.

The two sections and their constituent parts are described in the following paragraphs.

8.5.1.2 GMBS Multi-Mode Terminal Architecture

All the subsystems constituting the GMMT revolve around the T-IWU, which has three main interfaces towards the link environments, *viz.*: the wireless segments, the user, and the navigation system.

The wireless segment is based on the three radio environments:

- the M-ESW satellite segment;
- the GPRS segment;
- the W-LAN segment, which acts as an extension of the satellite segment.

The user side provides the interface to the Internet applications under mobility conditions.

Finally, of significant importance is the navigation system, providing the T-IWU and the satellite antenna system with the elements to track the satellite, taking into account the movement and the velocity of the GMMT.

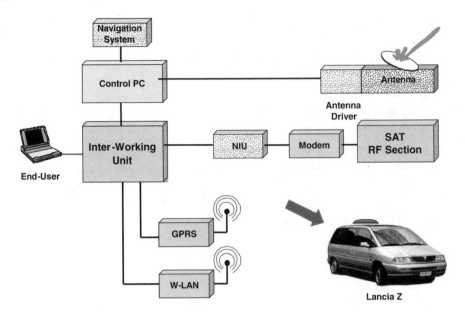

Figure 8.24   GMMT demonstrator block diagram.

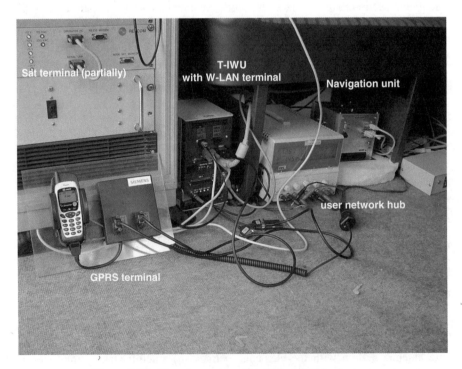

Figure 8.25   Experimental equipment installed in the van.

Figure 8.24 depicts the main blocks constituting the GMMT.

The practical implementation of the demonstrator block diagram is shown in Figure 8.25.

8.5.1.3   Fixed Network Segment Architecture

The fixed wireless segments correspond to the respective mobile access networks:

- a satellite access point, emulating the M-ESW satellite network, supporting all the IPMM capabilities;

- a GPRS/UMTS access point, based upon Base Station, SGSN (Serving GPRS Support Node) and GGSN (Gateway GSN) entities;

- a W-LAN access point, connected to a satellite gateway to furnish an extension to the M-ESW segment.

All the segments access the fixed network through an ER connected to a CR, both being located at a F-ISP.

A F-ISP is connected to other F-ISPs, each one having ER and CR capabilities.

8.5.1.4   Terminal Inter-Working Unit

The T-IWU manages handover, mobility and QoS of the mobile node. In particular, the T-IWU: (i) provides a local LAN to the user terminals; (ii) accesses the radio networks; (iii) manages handover; (iv) performs QoS functions; and (v) enables the support for inter-segment user mobility.

The T-IWU functionality is mainly based on routing functions. This allows single mobile nodes, as well as LANs consisting of several user terminals, to be connected to GMBS services. For both cases, an Ethernet LAN interface provides user access. The T-IWU also incorporates IPv4 to IPv6 conversion.

The T-IWU is additionally responsible for initiating and controlling the inter-segment handover. The T-IWU performs this function for the attached terminals by DHCP (Dynamic Host Configuration Protocol) stateless self-configuration. The route addresses of the T-IWU for each segment are obtained at call set-up every time a new segment is entered. For example, in the GPRS segment the IP address of the T-IWU GPRS port is obtained dynamically from the GGSN. Seamless IP handover is achieved by the use of Mobile IPv6 [IET-03].

In addition to the user data being sent or received through one of the segments, the T-IWU has to interface with each baseband section of the segment modems for call/session set-up and control (e.g. AT (Attention) commands, M-ESW call/session set-up and control commands). Moreover, monitoring of the actual segment parameters (general availability, receive link power, link quality) has to accessed by the T-IWU for the handover determination.

This requires two different interfaces for the IP data and the control/monitoring data. The control/monitoring data and command sets differ for each segment modem and have to be adopted for each segment.

8.5.1.5 Navigation Unit

The demonstrator makes use of the navigation unit to access data for the broadband Satellite Terminal (SaT) antenna steering. The antenna pointing, acquisition and tracking (PAT) of the SaT focuses on an open loop method by making use of position and attitude information of the mobile. The position requirements are relaxed, because relatively large position changes of the order of some hundreds of metres produce only small changes in the elevation angle.

The navigation unit consists of a commercially available car navigation unit and a two-axis angular sensor. The main parts of the car navigation unit are a GPS-receiver, some car information sensors and a digital map. Position and location are delivered according to the digital map. In addition to more accurate positioning by map matching, the advantage of the digital map is that the co-ordinates of the user environment, i.e. city or land, is known. This information can also be utilised for the support of handover procedures and for the support of mobile Internet applications.

Figure 8.26 shows the functional block diagram of the signal flows between the navigation unit and the different units of the GMMT.

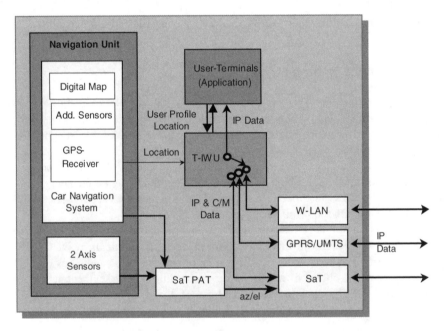

Figure 8.26 Navigation unit functional block diagram. Reproduced by permission from Kluwer Academic Press © 2003.

8.5.1.6 Satellite Terminal

The M-ESW SaT equipment is based on three main parts: the Terminal Network Interface Unit (T-NIU), a commercial satellite access modem with an IF section and the agile phased array antenna with a RF section.

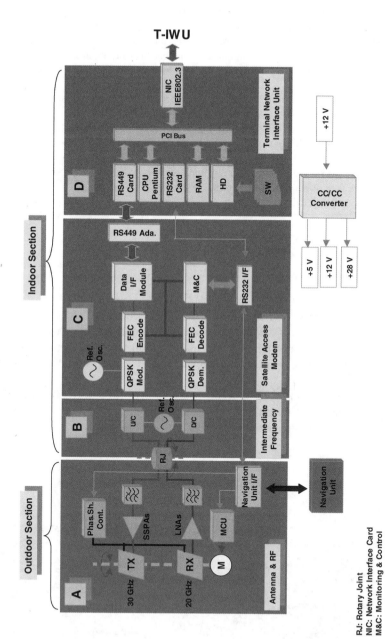

Figure 8.27 Broadband satellite terminal configuration.

RJ: Rotary Joint
NIC: Network Interface Card
M&C: Monitoring & Control

Table 8.7   Main parts of the M-ESW system.

Part	Description
[A]	This block constitutes the RF front-end. It is composed by an agile phased array antenna based on the transmission and the reception sections both accommodated on a metallic plate for the mechanical motion in the azimuth and hosting the Low Noise Amplifier (LNA) and Solid State Power Amplifier (SSPA) components suitable for working at 20/30 GHz frequencies. A suitable rotary joint links the antenna with the Intermediate Frequency (IF) section.
[B]	This part is the IF section based on dual stage up and down converters to translate the frequencies from 20 GHz to 70 MHz and from 70 MHz to 30 GHz.
[C]	The modulator and demodulator (modem) electronics constitute this section. The system is based on off-the-shelf equipment. The modem supports all the encoder/decoder functionalities with error correction capability.
[D]	The T-NIU implements on the satellite side the Wireless Adaptation Layer (WAL) requirements, constituting a transparent bridge between the T-IWU and the M-ESW access environment.

Figure 8.27 shows the wireless M-ESW subsystem on the GMMT.

Four main parts compose the M-ESW subsystem and are summarised in Table 8.7.

In Figure 8.27, the ancillary components constituted by the azimuth actuator of the antenna, and the low voltage power supply system, complete the general GMMT M-ESW subsystem description.

Figure 8.28 shows the SaT, deployed on a test bench to perform the overall laboratory tests before the integration in the GMMT. In the picture from left to right can be seen: the modem, the T-NIU and the RF plus the phased array antenna.

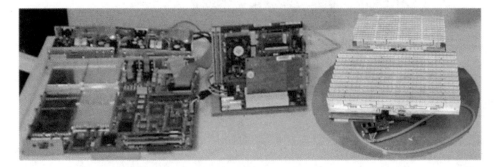

Figure 8.28   M-ESW satellite terminal demonstrator.

Terminal Network Interface Unit

The T-NIU constitutes the core subsystem to gain access to the M-ESW satellite segment. It connects the T-IWU with the satellite access modem by three different interfaces:

1. a RS232 interface connected towards the SaT modem to control/configure the device and to read the radio link parameter information (E_b/N_0 and Bit Error Rate).

2. a RS449 interface used to send and receive data (M-ESW cells) to and from the SaT modem.

3. an Ethernet IEEE 802.3 interface connecting the T-NIU with the T-IWU. Through this interface, IP packets are exchanged, carrying the data and the radio link ancillary information used by the T-IWU for operations/decisions.

The T-NIU was implemented using an embedded industrial Personal Computer.

Front-End

The M-ESW satellite segment is based on an M-ESW emulator system, as shown in Figure 8.29.

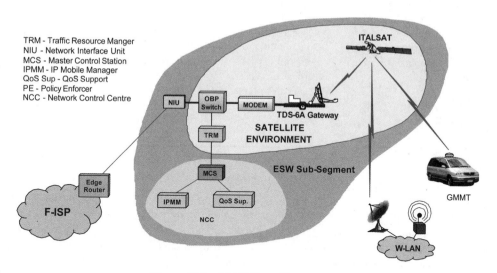

Figure 8.29 M-ESW satellite segment.

The satellite environment is divided into two parts:

- the actual satellite link, based on the ITALSAT Ka-band Italian satellite configured in transparent mode (National Global Coverage—NGC) [CAR-95];

- the switching payload based on an on-board processing switch, adapted to support the M-ESW emulation functionality.

The satellite is linked to the payload, physically located in a laboratory, through a satellite link station: the TDS-6/A (this will be discussed shortly).

The air interface towards the satellite represents a crucial part of the GMBS prototype. The technical requirements for the SaT are the dynamic pointing of the antenna beam to the satellite, and the implementation of a high gain, high EIRP (Effective Isotropic Radiated Power), compact Ka-band antenna.

Phased Array Antenna

GMBS foresees three terminal types, differing mainly in the antenna subsystem: a car-mounted terminal (utilising a non-protuberant flat antenna), a terminal for large vehicles (protuberant antenna) and a portable terminal; Table 8.8 shows the main requirements for the three experimental terminal typologies.

Table 8.8 GMBS terminal parameters

Type	Car-mounted	Protuberant	Portable
Uplink frequency (GHz)	29.747 ± 0.248	29.747 ± 0.248	29.747 ± 0.248
Downlink frequency (GHz)	19.950 ± 0.248	19.950 ± 0.248	19.950 ± 0.248
Uplink rate (max kbps)	256	512	512
Downlink rate (max kbps)	512	2048	2048
Antenna type	Planar Array	Dish	Dish
Antenna dimensions	20 cm × 40 cm	35 cm (radius)	35 cm (radius)
Pointing requirements	Open Loop	Open Loop	Open Loop
Beam steering	Electrical Elevation Mechanical Azimuth	Mechanical	Mechanical
Pointing error (deg)	5	3	3
G/T (dB/K)	3.1	6.8	6.8
EIRP (dBW)	37.3	41.1	40.2

The car-mounted terminal is the most critical because it requires the design of an innovative antenna that, while maintaining the correct radiometric parameters, satisfies the space constraints in a car: an agile phased-array antenna. The pointing system makes the antenna suitable to track the satellite position in a critical environment, taking into account the car's speed (more than 80 km/h) and the stability conditions on the road.

The Ka-band phased array satellite antenna is realised by active arrays, at 19 GHz for reception and at 29 GHz for transmission. The antenna beamwidth is about 2° in azimuth and 7° in elevation. For minimised antenna height, the arrays are mounted on a turntable and mechanically steered towards the satellite. Elevation beam agility is realised electronically.

The antenna system is connected to the UP/DOWN converter through a microwave rotary joint system, enabling continual revolution of n times 360°. The movable part of the antenna consists of the Ka-band front-end units with radiating elements (power amplifiers, LNAs, phase shifters, diplexer and printed dipoles sub-arrays) and is fully transparent for signals processed in different types of SaTs (UP/DOWN converters, modems etc). The fixed part of the antenna is connected to a control system (over RS232) and a power supply unit.

Figure 8.30   TDS-6 Trailer with the antenna deployed.

Satellite Gateway Description

The TDS-6 Station is a transportable type. Figure 8.30 shows the trailer with the antenna deployed.

The station is composed of three sections: a trailer, where the antenna and RF front-end are located; a shelter housing the base band equipment and the control user interface; and the cross-site cables to join the trailer and shelter equipment.

8.5.1.7   Space Segment Description: ITALSAT

The available satellite for the GMBS trial, as noted previously, was ITALSAT F2 in NGC mode configuration [CAR-95].

8.5.1.8   GPRS Segment

For the purpose of the demonstrator, a commercial GPRS mobile terminal was used. By the addition of the appropriate interface and drivers the terminal was able to fulfil the GMBS requirements. Part of this adaptation required a splitter element, allowing the connection of two serial links RS232 (CCITT/ITU-T V.24/V.28) to the T-IWU/GPRS interface. In the main serial link, the Data Link (IP Data transmission), a Point-to-Point Protocol session is set-up by AT (Attention) commands. In the other link, the Information Link (segment monitoring), the implemented binary protocol with specific commands provides the information requested by the T-IWU, simultaneously to an IP session.

8.5.1.9   W-LAN Satellite Extension

The W-LAN segment is based on a commercial product providing the practical implementation of the IEEE 802.11b standard. This equipment is based on two main parts: a base station, or access point, having a coverage range from 100 m to 300 m; and a MT suitable for desk top and notebook installation. In the GMBS demonstrator, the W-LAN was used in two possible configurations: as a bridge transponder to link the satellite and the user, in a particular indoor environment (e.g. inside buildings); and with a direct connection with the fixed network through an ER. The T-IWU is able to request link quality measurements from the W-LAN card for the purpose of segment selection.

8.5.2   Performance Evaluation

8.5.2.1   IP Mobility Support

A set of data and parameters was recorded to evaluate the performance of the mobility algorithms and the access segments.

In particular these measured parameters were:

- Link information for each segment (segment indicator, link parameters, power, $E_b/N_0, \ldots$)

- Attached User Terminal information (number, routing segment, profile)

- Mobility protocol information (Time of Binding Update Request and Acknowledgement for each User Terminal)

- QoS information of the segment (delay by Internet Control Message Protocol, lost packets)

- General information (GPS time and date, GPS position of the Mobile)

Figure 8.31 shows an evaluation of a 2.5-minute sequence of recorded data from the field trials. Here, in order to investigate the influence of user preference, the allocation of two users with different user profiles (business and value) is shown in comparison to the segment monitoring. Handover in the demonstrator realisation was based on several trigger conditions:

- the segment availability, as measured from the received power of the radio modems;

- the segment performance;

- the segment pricing;

- a service profile chosen by the users.

An operator-triggered handover can also be initiated.

In Figure 8.31(a) it should be noted that GPRS is always available during the recorded measurement period (100% time-share) and is therefore not illustrated in the figure. The availability of the satellite network is directly input (63.8% time-share), while for W-LAN (57.7% time-share) power measurements at the T-IWU are depicted (see also Table 8.9).

Figure 8.31(b) shows the traffic routing of the attached mobiles through the different access segments for users with different user profiles. The W-LAN segment is assumed to be the optimum and the best value, so both user types are tunnelled through the W-LAN, if

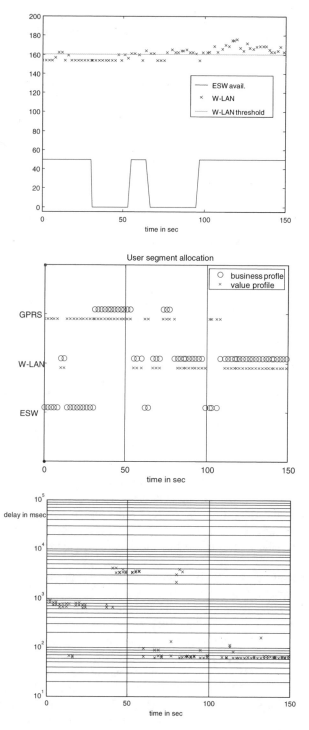

Figure 8.31 Results for (a) segment availability; (b) user segment allocation and (c) round trip delays.

Table 8.9 Segment availability for measurement period

Time-share of W-LAN availability	57.72%
Time-share of M-ESW availability	63.76%
Time-share of GPRS availability	100%

Table 8.10 Time-share of users with different profiles

Time-share of business user in segment	GPRS 20.81%	W-LAN 51.68%	M-ESW 27.51%
Time-share of value user in segment	GPRS 49.66%	W-LAN 50.34%	M-ESW 0%

Table 8.11 IP round trip delay to the home agent

Round trip time	Min	Mean	Max
W-LAN	62 ms	70 ms	161 ms
M-ESW	640 ms	747 ms	905 ms
GPRS	2131 ms	3468 ms	4100 ms

available. When W-LAN is not available, the route taken by the traffic depends on the user's profile, that is the business user is routed via the satellite segment, which offers the higher throughput rate and the value user via GPRS, which is the most economical. Table 8.10 shows the statistical time-share for the two user profiles in the three access segments. As GPRS is always available, it is obvious that mobiles with value user profiles do not use the satellite segment.

The measured statistics of the IP round trip delay to the home agent are shown in Table 8.11 and is illustrated in Figure 8.31(c) for a mobile user with a business profile, thus the delay values correspond with the circles in Figure 8.31(b).

It can be noted that the delay in the GPRS segment can exceed, by far, the delay of the satellite segment. In order to shed more light onto this fact, a routing trace for the GPRS segment is given in Table 8.12. Here, of interest, is especially the first hop from the GPRS Mobile Terminal to the GGSN which accounts for about 1.5 s of delay.

The QoS request of the PDP context activation during the trials is characterised by a Class 4 Delay, which means BE for any packet size. This is a current limitation in the GPRS access network, but fewer delays will be encountered with QoS requests for higher Classes becoming available. On the other hand, the Delay Classes for GPRS Services [3GP-03] include the delay for request and assignment of radio resources and consume the major part

Table 8.12 GPRS trace GPRS mobile terminal to ISP home agent

Hop number	IP address	Description	RTT ms
0	192.168.0.63	GPRS Mobile Terminal	0
1	192.168.1.5	GGSN	1465
2	192.168.140.43	ISDN (Integrated Services Digital Network) Router Vienna	1467
3	192.168.150.44	ISDN Router Vienna	1489
4	192.168.160.101	ISDN Router Rome	1748
5	192.168.160.100	ER Rome	1755
6	192.168.129.1	ISP Home Agent	2437

of this parameter. Some measurements in the presented scenario state delays of 10–12 seconds at this stage. There are no constraints for the upgrade stages (transit delays).

8.5.2.2 QoS Performance Evaluation

Each test has been carried out using access through each of the three wireless segments separately and considering, during the vehicle movement, for what concerns the W-LAN, two means of utilisation: as extension of the M-ESW satellite coverage where the satellite visibility is not available (e.g. indoor situations); as direct wireless access to the satellite access network, having the same configuration with respect to the GPRS and the M-ESW.

To describe the measurement layout it is necessary to refer to [ITU-99], where all the actors in the measurement layout are identified.

Figure 8.32 shows the GMBS demonstrator layout, highlighting the entities composing the system.

Direct measurements are performed at two levels:

- At IP packet level, generating the packet at the Source Host (SRC) on the Monitoring Point 1 (MP1), then monitoring the packet number and its arrival sequence at the Destination Host (DST) at the Measurement Point 2 (MP2);

- At time level, measuring the IP packet start time at the SRC on the MP1, the destination time at DST on MP2 and, finally, the RTT at the SRC on the MP1 when the IP packet returns.

Figures 8.33, 8.34 and 8.35 report the validation campaign results. Diagrams are grouped per access segment and for each group:

- The columns report the three QoS scenarios, i.e.: BE; DiffServ and DiffServ & GRIP;

- The rows report the IPTD, IPDV and the throughput measured for the three scenarios.

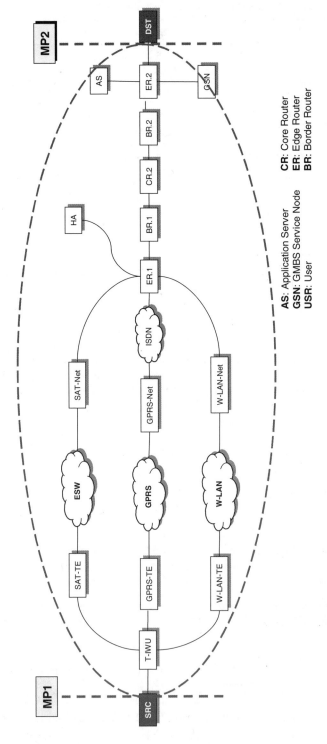

Figure 8.32 Measurement layout. Reproduced with permission from Kluwer Academic Press © 2003.

AS: Application Server
GSN: GMBS Service Node
USR: User

CR: Core Router
ER: Edge Router
BR: Border Router

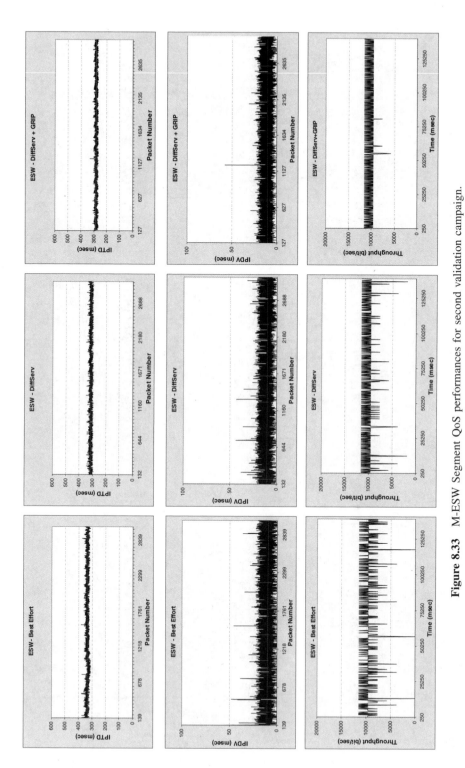

Figure 8.33 M-ESW Segment QoS performances for second validation campaign.

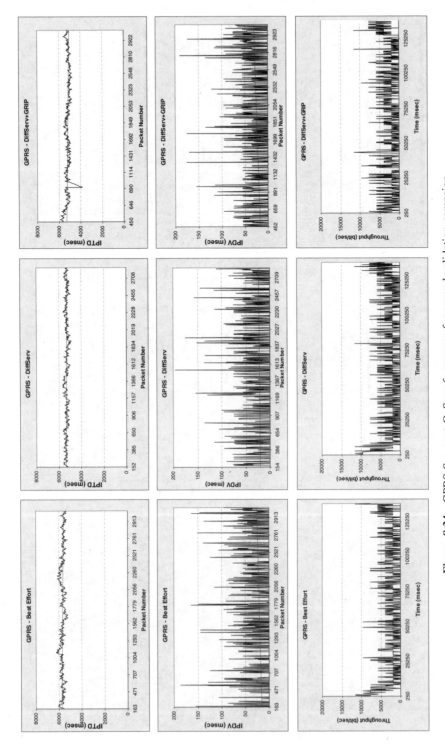

Figure 8.34 GPRS Segment QoS performances for second validation campaign.

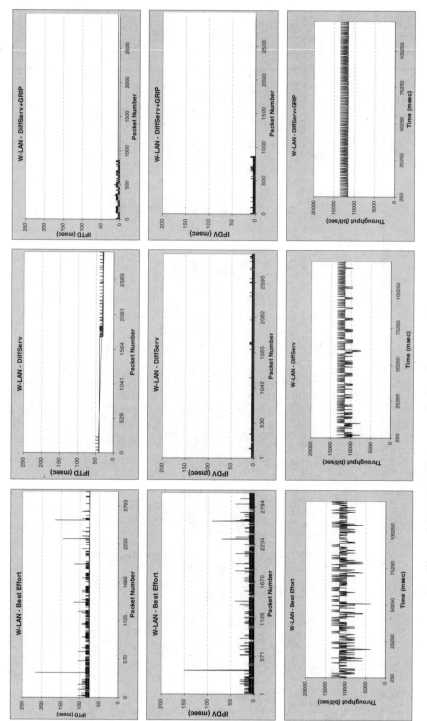

Figure 8.35 W-LAN Segment QoS performances for second validation campaign.

Table 8.13 Statistical results for second validation campaign

Access Segment	Traffic Configuration	Parameters										
		E2E delay (ms)				IPDV (ms)			Throughput (bit/s)		Packet loss ratio	
		Mean value	•	Min	Max	Mean value	•	Max	Mean value	•	Eval.	%
M-ESW Satellite Emul.	Best effort	335.400	10.019	309.500	383.000	10.601	6.743	57.500	9582.734	1656.175	0.076	7.57%
	DiffServ	314.012	9.338	292.000	349.000	10.337	6.513	43.500	10096.373	1334.922	0.024	2.43%
	DiffServ & GRIP	282.505	8.955	262.000	331.000	10.229	6.402	58.000	10283.506	969.101	0.003	0.27%
GPRS	Best effort	5691.682	198.153	5080.500	5723.000	60.467	40.210	175.000	2378.300	2386.035	0.773	77.26%
	DiffServ	5307.678	148.558	5000.500	5629.000	57.015	30.306	183.000	2623.538	2616.029	0.748	74.84%
	DiffServ & GRIP	5140.607	258.359	4737.500	5754.500	54.412	29.656	180.000	2374.828	2437.916	0.773	77.25%
W-LAN (IEEE 802.11b)	Best effort	80.214	9.815	75.000	160.500	4.984	8.868	151.000	10681.242	1693.134	0.105	10.50%
	DiffServ	39.655	1.116	30.000	55.000	0.289	1.334	10.500	11623.784	1177.695	0.030	3.00%
	DiffServ & GRIP	4.509	0.500	0.000	15.000	1.000	0.000	10.500	11983.035	822.211	0.000	0.00%

Table 8.13 reports the statistical results.

A large difference between the performances of the three access segments can be observed, the worst case is represented by the GPRS, that has not only a reduced data bandwidth but also a large packet loss and IPDV (see Table 8.13). This characteristic has been verified during the validation campaign at two sites: (i) where there was direct access to the GPRS network and (ii) where commercial access was used.

For what concerns the W-LAN, in Table 8.13 and in the diagrams, the data acquired have been obtained in a configuration where the W-LAN is directly connected to the M-ESW payload.

Finally, the M-ESW satellite access segment has demonstrated good reliability, as in the case of W-LAN, showing in the data the known access delay of the geostationary satellite, in fact the mean value of the IPDV is about 270 ms, considering the traffic configuration (see Table 8.13). Another important aspect evaluated during the validation campaigns is the large amount of time taken-up by the M-ESW access set-up (more that 4.5 s).

The validation campaign has shown that:

1. The BE solution produces very high packet losses and transfer delays when the network is congested; this is critical for the utilisation of the GMBS access with time-sensitive (e.g. VoIP, videoconference, etc.) applications and services.

2. The DiffServ solution improves the network performances with respect to the BE when the DiffServ classes are not congested, but the performance decreases quickly when the congestion increases (independently from the class considered); so also this solution is not suitable for executing time-sensitive applications in the network.

3. The GRIP mechanism provides good performances also when the network is congested and it can be configured to satisfy the requirements of time-sensitive applications. In this frame, it is important to observe that the CAC of the GRIP solution, adopted during the validation campaign, has been configured to maximise the network utilisation and not the network performances; changing the configuration, the GRIP solution optimises the IPTD and the IPDV.

4. The positive effect of GRIP on performance is more evident in the W-LAN segment in comparison to those of the ESW and GPRS. This is because in the latter cases the weight of congestion phenomena on overall performance is smaller. Since, obviously, an admission control function like GRIP controls only congestion effects, it cannot counter performance impairments due to other causes (e.g., propagation delay).

REFERENCES

[3GP-03] 3GPP TS 22.060 V6.0.0 (2003) 3rd Generation Partnership Project; Technical Specification Group Services and Systems Aspects; General Packet Radio Service (GPRS); Service description, Stage 1 (Release 6).

[BIA-02] G. Bianchi, N. Blefari-Melazzi, M. Femminella: Per-Flow QoS over a Stateless Differentiated Services IP Domain, *Computer Networks—The International Journal of Computer and Telecommunications Networking*, **40**(1), 16 September 2002; 73–87.

[CAR-95] F. Carducci, M. Francesi: The Italsat Satellite System, *International Journal of Satellite Communications*, **13**(1), January-February 1995; 49–81.

[IET-03] D.B. Johnson, C. Perkins, J. Arkko: Mobility Support in IPv6, *Internet Engineering Task Force*, Internet Draft draft-ietf-mobileip-ipv6-24.txt, Work in Progress, 30 June 2003.

[IET-99] J. Heinanen, F. Baker, W. Weiss, J. Wroclawski: Assured Forwarding PHB Group, *Internet Engineering Task Force*, RFC 2597, June 1999.

[ITU-99] ITU-T Recommendation I.380, Internet Protocol Data Communication Service—IP Packet Transfer and Availability Performance Parameters, (02/99).

ACRONYMS

AF	Assured Forwarding
AS	Application Server
AT	Attention
BE	Best Effort
BR	Border Router
CAC	Connection Admission Control
CR	Core Router
DHCP	Dynamic Host Configuration Protocol
DiffServ	Differentiated Services
DST	Destination Host
E2E	End-to-End
EF	Expedited Forwarding
EIRP	Effective Isotropic Radiated Power
ER	Edge Router
ESW	EuroSkyWay
F-ISP	Federated ISP
FTP	File Transfer Protocol
GGSN	Gateway GPRS Support Node
GMBS	Global Mobile Broadband System
GMMT	GMBS Multi-Mode Terminal
GPRS	General Packet Radio Service
GPS	Global Positioning System
GRIP	Gauge&Gate Reservation with Independent Probing
HTTP	Hyper Text Transfer Protocol
IETF	Internet Engineering Task Force
IF	Intermediate Frequency
IntServ	Integrated Services
IP	Internet Protocol
IPDV	IP Packet Delay Variation
IPMM	IP Mobility Management
IPTD	IP Packet Transfer Delay
IPv4/6	IP version 4/6
ISDN	Integrated Services Digital Network
ISP	Internet Service Provider
LAN	Local Area Network

LNA	Low Noise Amplifier
MAC	Medium Access Control
MCS	Master Control Station
M-ESW	Mobile EuroSkyWay
MP	Measurement/Monitoring Point
MSMT	Multi-Segment Mobile Terminal
MT	Mobile Terminal
NGC	National Global Coverage
NIU	Network Interface Unit
NOC	Network Operation Centre
OS	Operating System
PAT	Pointing, Acquisition and Tracking
PDCP	Packet Data Convergence Protocol
PDP	Packet Data Protocol
PE	Policy Enforcer
QoS	Quality of Service
RF	Radio Frequency
RLC	Radio Link Control
RSVP	Resource Reservation Protocol
RTT	Round Trip Time
SaT	Satellite Terminal
SGSN	Serving GPRS Support Node
SLA	Service Level Agreement
SRC	Source Host
SSPA	Solid State Power Amplifier
T-IWU	Terminal Inter-working Unit
T-NIU	Terminal Network Interface Unit
TRM	Traffic Resource Manager
UMTS	Universal Mobile Telecommunications System
USR	User
VoIP	Voice over IP
WAL	Wireless Adaptation Layer
W-LAN	Wireless Local Area Network

Appendix A
Related Publications

P. Conforto, C. Tocci, V. Schena, L. Secondiani, N. Blefari-Melazzi, P.M.L. Chan, F. Delli Priscoli: End-to-End QoS and Global Mobility Management in an Integrated Satellite/Terrestrial Network, *International Journal of Satellite Communications and Networks*, **22**(1), January–February 2004; 19–53.

G. Bianchi, N. Blefari-Melazzi, P.M.L. Chan, M. Holzbock, Y.F. Hu, A. Jahn, R.E. Sheriff: Design and Validation of QoS Aware Mobile Internet Access Procedures for Heterogeneous Networks, *Journal on Special Topics in Mobile Networking and Applications (MONET)—Special issue on Personal Environment Mobility* **8**(1), February 2003; 11–25.

V. Schena, G. Losquadro: Wideband Ka Satellite Terminal Design: Prototype Implementation and on the Field Tests, *Proceedings of 8th Ka Band Utilization Conference*, 25–27 September 2002, Baveno, Italy; 145–152.

G. Bianchi, N. Blefari-Melazzi, M. Femminella: Per-flow QoS Support over a Stateless DiffServ Domain, *Computer Networks: The International Journal of Computer and Telecommunications Networking—Special Issue on The New Internet Architecture*, **40**(1), 16 September 2002; 73–87.

M. Holzbock, A. Jahn, J. Alonso, Z. Golubicic, V. Schena, F. Ceprani: SUITED Demonstration Results of a Mobile Terminal for Heterogeneous Satellite-Terrestrial IP Network Access, *Proceedings of IST Mobile and Wireless Telecommunications Summit 2002*, 16–19 June, 2002 Thessaloniki, Greece; 90–94.

V. Schena, F. Ceprani, F. Vecchia: SUITED Demonstrator Validation Campaign: First Results on the End-to-End IP QoS Measurement, *Proceedings of IST Mobile and Wireless Telecommunications Summit 2002*, 16–19 June, 2002 Thessaloniki, Greece; 394–398.

L. Secondiani, F. Mazzolini, N. Blefari-Melazzi, M. Femminella, L. Piacentini: A Transport Interworking Protocol for the Support of TCP/IP based Applications over the ESW Satellite Network in the GMBS Environment, *Proceedings of IST Mobile and Wireless Telecommunications Summit 2002*, 16–19 June, 2002 Thessaloniki, Greece; 773–777.

P. Conforto, C. Tocci, N. Blefari-Melazzi, M. Femminella, P. Dini, R. De Finis: QoS-Aware Mobility Management: Solutions and Performance Evaluation, *Proceedings of IST Mobile and Wireless Telecommunications Summit 2002*, 16–19 June, 2002 Thessaloniki, Greece; 483–488.

P.M.L. Chan, R.A. Wyatt-Millington, A. Svigelj, R.E. Sheriff, Y.F. Hu, P. Conforto, C. Tocci: Performance Analysis of Mobility Procedures in a Hybrid Space Terrestrial IP Environment, *Computer Networks: The International Journal of Computer and Telecommunications Networking, Special Issue Broadband Satellite Systems: A Networking Perspective, Elsevier Science B.V,* **39**(1), 15 May 2002; 21–41.

P.M.L. Chan, R.E. Sheriff, Y.F. Hu, P. Conforto, C. Tocci: Design and Evaluation of Signaling Protocols for Mobility Management in an Integrated IP Environment, *Computer Networks: The International Journal of Computer and Telecommunications Networking, Elsevier Science B.V,* **38**(4), 15 March 2002; 517–530.

Space/Terrestrial Mobile Networks. Edited by R.E. Sheriff, Y.F. Hu, G. Losquadro, P. Conforto, C. Tocci
© 2004 John Wiley & Sons, Ltd ISBN: 0-470-85031-0

P.M.L. Chan, R.E. Sheriff, Y.F. Hu: Implementation of Fuzzy Multiple Objective Decision Making Algorithm in a Heterogeneous Mobile Environment, *Proceedings of WCNC 2002: IEEE Wireless Communications and Networking Conference*, **1&2**, 17–21 March 2002, Orlando, USA; 332–336.

P.M.L. Chan, Y.F. Hu, R.E. Sheriff: Fuzzy Handover for Heterogeneous Mobile Networks, *Proceedings of The 20th IASTED International Multi-Conference Applied Informatics (AI 2002)*, 18–21 February, 2002, Innsbruck, Austria; 243–248.

P. Conforto, C. Tocci, G. Losquadro, R.E. Sheriff, P.M.L. Chan, Y.F. Hu: Ubiquitous Internet in an Integrated Satellite-Terrestrial Environment: the SUITED Solution, *IEEE Communication Magazine Special Issue on Service Portability and Virtual Home Environment*, **40**(1), January 2002; 98–107.

P.M.L Chan, R.E. Sheriff, Y.F. Hu, P. Conforto, C. Tocci: Mobility Management Incorporating Fuzzy Logic for a Heterogeneous IP Environment, *IEEE Communications—Focus Issue on Evolving to Seamless All IP Wireless/Mobile Networks*, **39**(12), December 2001; 42–51.

G. Bianchi, N. Blefari-Melazzi, M. Femminella, F. Pugini: Performance Evaluation of a Measurement-Based Algorithm for Distributed Admission Control in a DiffServ Framework, *Proceedings of IEEE Global Telecommunications Conference (Globecom 01): Internet Performance Symposium*, 25–29 November 2001, San Antonio, Texas, USA; 1886–1891.

G. Bianchi, N. Blefari-Melazzi: Admission Control over Assured Forwarding PHBs: a Way to Provide Service Accuracy in a DiffServ Framework, *Proceedings of Global Telecommunications Conference (Globecom 01) Conference: Quality of Service in Computer Networks Symposium*, 25–29 November 2001, San Antonio, Texas, USA; 2561–2565.

C. Tocci, P. Conforto, F. Fedi, G. Losquadro: Delivery of QoS Sensitive Mobile Services over an IP-based Satellite Network Complemented with Wireless Terrestrial Components, *Proceedings. of the 7th Ka Band Utilization Conference*, 26–28 September 2001, S. Margherita Ligure (GE), Italy; 503–510.

M. Holzbock, A. Jahn, V. Schena, F. Ceprani, M. Reale, V. Angarola: Mobility and QoS Support in the SUITED Multi-Segment IP Infrastructure, *Proceedings of the IST Mobile Communications Summit 2001*, 9–12 September 2001, Barcelona, Spain; 690–695.

J. Alonso, J. Perez, L. González, V. Schena, F. Ceprani, M. Holzbock: SUITED Vehicular Ka Band Satellite Terminal, *Proceedings of the IST Mobile Communications Summit 2001*, 9–12 September 2001, Barcelona, Spain; 120–125.

F. Delli Priscoli, G. Lombardi, F. Del Sorbo, D. Prisco, A. Fazio: Integrated Services Mappings over a Multi-Segment Broadband Wireless Network, *Proceedings of the IST Mobile Communications Summit 2001*, 9–12 September 2001, Barcelona, Spain; 816–824.

V. Marziale, A. Vitaletti: A Framework for Internet QoS Requirements Definition and Evaluation: an Experimental Approach, *Proceedings of the IST Mobile Communications Summit 2001*, 9–12 September 2001, Barcelona, Spain; 306–310.

N. Blefari-Melazzi, D. Di Sorte, M. Femminella, G. Reali: Resource Allocation Rules to Provide QoS Guarantees to Traffic Aggregates in a DiffServ Environment, *Proceedings of the IST Mobile Communications Summit 2001*, 9–12 September 2001, Barcelona, Spain; 810–815.

P. Conforto, C. Tocci, G. Losquadro, F. Fedi, R.E. Sheriff, A. Vitaletti: SUITED/GMBS System: Architecture and Mobile Terminal, *Proceedings of the IST Mobile Communications Summit 2001*, 9–12 September 2001, Barcelona, Spain; 126–132.

C. Tocci, P. Conforto, A. De Carolis, M. Femminella, F. Pugini: Solutions and Techniques for the Provision of End-to-End IP QoS in a GMBS System, *Proceedings of the IST Mobile Communications Summit 2001*, 9–12 September 2001, Barcelona, Spain; 312–317.

P.M.L. Chan, R.E. Sheriff, Y.F. Hu: An Intelligent Handover Strategy for a Multi-Segment Broadband Network, *Proceedings. of 12th IEEE International Symposium on Personal, Indoor and Mobile Radio Communications, PIMRC 2001*, **1&2**; C55–C59.

C. Tocci, P. Conforto, G. Losquadro, N. Blefari-Melazzi, A. Fazio: Architectures and Protocols for the Provision of Ubiquitous IP-based QoS Aware Services over a Satellite and Terrestrial Complemented

System, *Proceedings of the 3rd Generation Infrastructure and Service Conference*, 2–3 July 2001, Athens, Greece; 74–78.

N. Blefari-Melazzi, D. Di Sorte, G. Reali: A Scalable CAC Technique to Provide QoS Guarantees in a Cascade of IP Routers, *IEEE International Conference on Communications*, 11–15 June, 2001, Helsinki, Finland; 654–658.

P. Conforto, G. Losquadro, V. Schena, C. Tocci: Broadband IP Mobile Service: Network and Technologies Developed in the Frame of the SUITED Project, *Proceedings of the 19th AIAA International Communications Satellite Systems Conference and Exhibit*, 17–20 April 2001, Toulouse, France.

M. Luglio, W. Pietroni: The Use of Hybrid Orbit Satellite Constellations for High Capacity Communications, *Proceedings of the 19th AIAA International Communications Satellite Systems Conference and Exhibit*, 17–20 April 2001, Toulouse, France.

A. Jahn, M. Holzbock: Mobile Multimedia IP Service Demonstration in Multi-Segment Terrestrial/ Satellite Networks, *Proceedings of the 19th AIAA International Communications Satellite Systems Conference and Exhibit*, 17–20 April 2001, Toulouse, France.

P. Conforto, C. Tocci, G. Losquadro, R.E. Sheriff, P.M.L. Chan: Global Mobility and QoS Provision for Internet Services: The SUITED Solution, *Proceedings of IEEE Global Telecommunication Conference (GLOBECOM'00) Workshop CFP: Service Portability and Virtual Customer Environments*, 27 November–1 December 2000, San Francisco, USA; 3–12.

N. Blefari-Melazzi, G. Reali: Improving the Efficiency of Circuit-Switched Satellite Networks by Means of Dynamic Bandwidth Allocation Capabilities, *IEEE Journal on Selected Areas in Communications*, **18**(11), November 2000; 2373–2384.

R.A. Wyatt-Millington, R.E. Sheriff, Y.F. Hu, P. Conforto, G. Losquadro: The SUITED Project: A Multi-Segment System for Broadband Access to Internet Services, *IEE Colloquium on Broadband Satellite: The Critical Success Factors*, 16–17 October 2000, London UK; 12/1–12/11.

G. Bianchi, N. Blefari-Melazzi, M. Holzbock, A. Jahn: QoS and Mobility Support in a Multi-Segment IP Infrastructure, *Proceedings of IST Mobile Communications Summit*, 1–4 October 2000, Galway, Ireland; 323–328.

P. Conforto, G. Losquadro, C. Tocci, N. Blefari-Melazzi: Mobility Management in the SUITED/GMBS Multi-Segment System, *Proceedings of IST Mobile Communications Summit*, 1–4 October 2000, Galway, Ireland; 601–607.

C. Tocci, P. Conforto, G. Losquadro, F. Delli Priscoli, M. Femminnella: Internet End-to-End QoS Support in a GMBS System: The SUITED Solution, *Proceedings of IST Mobile Communications Summit*, 1–4 October 2000, Galway, Ireland; 453–458.

M. Holzbock, A. Jahn, F. Grassl, J. van Noten, M. Pugliese, V. Schena, S. Dragas: Ka band Satellite Network Technology: Demonstration of Mobile Services in Multi Segment IP Networks, *Proceedings of IST Mobile Communications Summit*, 1–4 October 2000, Galway, Ireland; 663–668.

V. Schena, G. Losquadro, M. Holzbock, A. Jahn, J. Alonso, S. Dragas: SUITED Demonstrator Proposal for a New Technology in the Radio Frequency Front-End in the Ka band Satellite Terminals for Mobile Broadband Applications, *Proceedings of IST Mobile Communications Summit*, 1–4 October 2000, Galway, Ireland; 747–752.

F. Delli Priscoli, L. Falò: SUITED Architecture in a Fully UMTS Scenario, *Proceedings of IST Mobile Communications Summit*, 1–4 October 2000, Galway, Ireland; 237–242.

P. Conforto, G. Losquadro, C. Tocci, M. Luglio, R. E. Sheriff: SUITED/GMBS System Architecture, *Proceedings of IST Mobile Communications Summit*, 1–4 October 2000, Galway, Ireland; 115–121.

E. Sereni, G. Baruffa, F. Frescura: A Software Radio Architecture for Implementation Complexity Evaluation of Multi-Standard Transceivers, *Proceedings of IST Mobile Communications Summit*, 1–4 October 2000, Galway, Ireland; 267–272.

P. Conforto, C. Tocci, G. Losquadro: A Mobility Management Scheme for Global Mobile Broadband System Supporting Ubiquitous QoS Sensitive Internet Services, *Proceedings of 4th European*

Workshop on Mobile and Personal Satellite Communication, 18 September 2000, London, UK; 249–255.

F. Di Cola, P.M.L. Chan, R.E. Sheriff, Y.F Hu: Handover and QoS Support in Multi-Segment Broadband Networks, *Proceedings of 4th European Workshop on Mobile and Personal Satellite Communication*, 18 September 2000, London, UK; 92–101.

M. Holzbock, O. Lücke, B. Oeste: Mobility Requirements of Landmobile Antennas for Broadband Satellite Communication, *Proceedings of 4th European Workshop on Mobile and Personal Satellite Communication*, 18 September 2000, London, UK; 34–43.

C. Tocci, G. Losquadro, F. Fedi, P. Conforto: Internet End-to-End QoS Provisioning over Ka Band Satellite System in the GMBS Environment, *Proceedings of Sixth Ka Band Utilization Conference*, 31 May–2 June 2000, Cleveland, Ohio, USA; 285–292.

G. Losquadro: Global Integrated System for Future Broadband Mobile Satellite Communications, *Collection of the 18th AIAA International Communications Satellite Systems Conference and Exhibit, Technical Papers*, **1&2**, 10–14 April 2000, Oakland, CA, USA; 163–171.

Index

Space/Terrestrial Mobile Networks. Edited by R.E. Sheriff, Y.F. Hu, G. Losquadro, P. Conforto, C. Tocci
© 2004 John Wiley & Sons, Ltd ISBN: 0-470-85031-0